Thomas Jefferson and his Decimals 1775–1810: Neglected Years in the History of U.S. School Mathematics

M. A. (Ken) Clements • Nerida F. Ellerton

Thomas Jefferson and his Decimals 1775–1810: Neglected Years in the History of U.S. School Mathematics

Foreword by Douglas L. Wilson

 Springer

M. A. (Ken) Clements
Department of Mathematics
Illinois State University
Normal, IL, USA

Nerida F. Ellerton
Department of Mathematics
Illinois State University
Normal, IL, USA

ISBN 978-3-319-34710-3   ISBN 978-3-319-02505-6 (eBook)
DOI 10.1007/978-3-319-02505-6
Springer Cham Heidelberg New York Dordrecht London

Printed on acid-free paper

Springer is part of Springer Science+Business Media (www.springer.com)

Foreword

Especially for Americans, there is no escaping the influence of Thomas Jefferson. His imprint on things American is everywhere, even on the landscape. Observant air travellers flying from New York to Chicago, for example, will notice that the visible markings on the terrain become noticeably more rectilinear after crossing Pennsylvania. The reason is that from the state of Ohio westward the land was surveyed and laid out on a grid before settlement and sold or given out in rectangular plots. Lines created by this grid are still visible as roads or as the boundaries of fields, farms, towns, counties, and even states. This scheme was yet another brainchild of Thomas Jefferson, whose ideas and reforms issued forth in profusion during the early years of the American Republic.

Jefferson is best known, of course, as the author of the Declaration of Independence. Memorable for its statement of colonial grievances that constituted the rationale for the American Revolution, its bold affirmations of human rights and democratic values have made it a landmark of modern history. But the list of other memorable things Jefferson wrote or devised or promoted is long. He himself called attention to two of these, in addition to his authorship of the Declaration, in prescribing the inscription for his gravestone: his authorship of the Virginia statute for religious freedom, and the founding of the University of Virginia. Nothing about being elected Governor of Virginia or President of the United States or being responsible for the Louisiana Purchase, any one of which would have been sufficient to mark a truly notable career.

As an enthusiastic child of the Enlightenment, Jefferson concentrated on reforms that would contribute towards an educated citizenry and thus enhance self-government. To be successful, a democratic government needed an informed public. This required basic literacy and numeracy, which is why Jefferson pioneered as an advocate for common schools at public expense. His original plan actually provided for basic education but also for advanced training for the most talented students, who would then become the teachers and professionals needed to carry on the educational system and occupy the learned professions. The book in hand focuses on one of the many little-known reforms in this vein that Jefferson originated, namely, instituting a decimal system for money, as well as weights and measures. The traditional British units in both of these areas were notoriously arbitrary and not easily expressed arithmetically, making them burdensome to learn and use, and thus barriers to personal advancement. By contrast, a decimal system had the advantage not only of having a rationalized basis but of being much easier to learn.

But as the authors of this work make clear, in this particular reform Jefferson had to settle for half a loaf. While eminently successful in replacing the monetary system of pounds, shillings, and pence, his reforms in the fields of weights and measures were defeated. While some aspects of the existing systems seemed reasonable enough—a foot being comprised of 12 smaller units, for example—having 5,280 feet comprise a mile seems wholly arbitrary, and difficult to recall and employ. A teaspoon seems a reasonable measure to reckon with in certain situations, but that 768 teaspoons should make a gallon seems perverse. This would appear in retrospect to have been an especially vulnerable system for rationalization and reform, but except for scientific endeavors, it is largely still with us. The general adoption of the metric system has, in fact, been urged repeatedly, and there have been times when it appeared that America was on the verge of adopting it. In the 1970s, an American president,

Gerald Ford, even boasted that the United States was "miles ahead" in its adoption of the metric system, a characterization that ironically signaled the eventual failure of a scheme in which *miles* were to be replaced by *kilometers*.

The authors of this book make a notable contribution to Jefferson scholarship in a way that reminds us that the scholarly enterprise is a two-way street. Having drawn on the existing accumulation of what is known about Jefferson, they have substantially added to this store of knowledge by the application of their particular area of scholarly interest and expertise—the various ways that mathematics has been taught and learned in the past. Also remarkable in this regard is that this illuminating work on Jefferson comes on the heels of another work by the same authors that sheds new light on another subject of great interest— the education of Abraham Lincoln. In this instance, they have employed their highly specialized historical knowledge of cyphering books in mathematical education to interpret the earliest extant manuscripts in Lincoln's hand, which, properly understood, constitute what remains of his own cyphering book. In parallel to what they have done here for Thomas Jefferson, their distinctive background and experience has enabled the authors to contribute very significantly to our existing knowledge of Abraham Lincoln. An obvious lesson to be drawn from such developments is that what we call knowledge is not a static entity but rather an on-going process. And once again it is shown that in spite of the vast resources of time and effort that have been devoted to the study of these two historical giants, there is always more to learn.

Galesburg, IL   Douglas L. Wilson

Contents

List of Figures

List of Tables

List of Tables

Overall Book Abstract, and Individual Abstracts
for the Seven Chapters

Overall Abstract

This book tells how decimal fractions moved from a place on the periphery of school mathematics in North America towards a more central place. The stimulus for this development was the introduction of an official decimalized Federal form of money, largely as a result of the efforts of Thomas Jefferson. The analysis establishes the fact that Jefferson believed that if the fledgling U.S. government would introduce a decimalized national currency, as well as a decimalized coordinated system of weights and measures, then this would greatly assist ordinary people to acquire the numerical skills that they would need to survive with dignity in most aspects of daily living in the United States of America. Although Jefferson's efforts to achieve decimal currency were successful, he was not able to overcome the vested interests of those who opposed the introduction of a coordinated, decimalized system of weights and measures. Some consequences of this failure are discussed in this book.

The following five research questions provided the foci for the study:

1. To what extent did policies and practices with respect to money influence school arithmetic curricula in the North American colonies before 1792?
2. To what extent did policies and practices with respect to weights and measures influence school arithmetic curricula in the United States of America before 1792?
3. To what extent did the place of decimal fractions in U.S. school arithmetic curricula change as a result of the introduction of a decimalized scheme of Federal money in 1792?
4. To what extent did policies and practices with respect to weights and measures play a part in school arithmetic curricula in the United States of America between 1792 and 1860?
5. What are the implications of the findings of this study for mathematics curriculum theory?

These questions are examined through the lens of a lag-time theoretical perspective which has been developed by the authors.

It was found that before 1792 school arithmetic in North America was largely controlled by a long-established cyphering tradition by which boys (and some girls), after they had reached the age of 10, prepared handwritten cyphering books in which the topics covered were mainly concerned with the four operations of arithmetic and with money and weights and measures. After the U.S. Congress approved the introduction of a decimalized form of Federal money, in the mid-1780s, and the national Mint was opened in 1792, students began to learn to carry out money calculations in eagles, dollars, cents, dimes and mills. But they also continued to learn how to make calculations with sterling currency denominations (pounds, shillings, pence and farthings) because, until well into the second half of the nineteenth century, sterling currency was still legal tender in the United States of America.

Analysis of cyphering-book data and of school textbooks used in North America between 1700 and 1810 revealed that although author-intended arithmetic curricula—as found in textbooks—in the United States of America, always included vulgar fractions and decimal fractions, entries on these topics were usually conspicuous by their absence in the teacher-implemented arithmetic curricula (as identified by analyzing cyphering books prepared by students). However, gradually, after 1792, a greater proportion of U.S. cyphering books included entries on decimal fractions. That gradual acceptance of decimal fractions is consistent with a lag-time theoretical position which predicted that, as a result of Jefferson's success in establishing a national, decimalized form of currency, schools were virtually compelled to teach their students about decimal currency—and from that situation it was a short step to getting students to study decimal fractions.

In Chapter 6, a comparative analysis of teacher-implemented arithmetic curricula in North America and Great Britain during the eighteenth and much of the nineteenth centuries is presented. This analysis led to the conclusion that, whereas in Great Britain there was no progress toward getting proportionally more students to study decimal fractions, in the United States, there was a steady growth in the proportion of students studying decimal fractions. That finding was consistent with predictions from the lag-time theoretical position.

From a mathematics education curricular perspective, perhaps the most important finding of the study reported in this book is that often there was a gulf between author-intended and teacher-intended arithmetic curricula. These large differences between the two types of curricula, which seem to have escaped the attention of previous historians, justify the claim, implicit in this book's subtitle, that the period 1775–1810 has been neglected so far as the history of school mathematics in the United States of America is concerned.

Individual Chapter Abstracts

Chapter 1: 1776: Dawn of a New Day in School Mathematics in North America

Abstract: The 1776 Declaration of Independence announced the dawn of a new day, for North America and, indeed, for the world. But did it herald new a new day for school mathematics? After describing how the term "school" was used in 1776 it is pointed out that at that time only a small proportion of North American adults had studied arithmetic beyond the four operations. The chapter examines the views of Thomas Jefferson on school education, in general, and on school mathematics, in particular. It is argued that it was within Thomas Jefferson's genius that a key to transforming school mathematics in the United States of America would be found.

Chapter 2: The *Abbaco* Curriculum in Colonial Schools in North America Before 1776

Abstract: This chapter examines the extent to which arithmetic curricula in schools in North America before 1776 were concerned with calculations and problems involving money and weights and measures. Analyses of chapters in arithmetic textbooks (deemed to be "author-intended curricula") and handwritten entries in students' cyphering books (regarded as evidence of "teacher-implemented arithmetic curricula") reveal that tasks related to money and to weights and measures permeated what historians have called the *abbaco* curriculum. That curriculum influenced the way the content of arithmetic was structured and sequenced in education institutions below the college level. The analyses also reveal that although money tasks and weights and measures tasks provided the principal emphases in school arithmetic, many students merely applied rules and methods which they did not understand.

Chapter 3: Thomas Jefferson and an Arithmetic for the People

Abstract: This chapter focuses on U.S. currency issues, and policies, practices and reforms during the period 1775–1792. As a result of post-war agreements spelled out by the Treaty of Paris (1783), the fledgling U.S. Congress needed to establish an official currency for the new nation. At that time, the most powerful financial figure in the nation was Robert Morris, Superintendent of Finance between 1781 and 1784, but aspects of his proposal for a new currency system were problematized by the young, and influential Thomas Jefferson, already a former Governor of Virginia, and famous for having drafted the Declaration of Independence. Like Morris, Jefferson proposed a decimal-based system of coinage, but the units for Morris's and Jefferson's systems were different. It was Jefferson's system which prevailed, and the most startling thing about his success on this matter was that his fundamental argument belonged to the realm of mathematics education—a combination of mathematics and education. Jefferson argued that his system would assist all U.S. citizens to achieve a better grasp of basic arithmetic than ever before, and that that would make it easier for them to survive with dignity. This chapter summarizes Jefferson's main arguments for the introduction of his version of decimal currency, and why he thought educational issues were so important.

Chapter 4: Weights and Measures in Teacher-Implemented Arithmetic Curricula in Eighteenth-Century North American Schools

Abstract: The decision having been made by the U.S. Congress to establish a decimal system of currency as the official national currency of the United States of America, one might have expected a decimal system of weights and measures to follow quickly. But that was not to be. Between 1784 and 1789 Thomas Jefferson was in France, and therefore his role as principal catalyst for decisions on decimalization was muted. The chapter begins by placing the weights and measures decision in the context of economic and political forces operating within the new nation. From a lag-time theoretical perspective, although there were numerous available arithmetic textbooks which dealt with decimal fractions, the prevailing economic and political pressures negated any educational and mathematical pressures for greater use of decimals. Thus, a curious result occurred—the United States decimalized its currency, but not its weights and measures.

Chapter 5: Decimal Fractions and Federal Money in School Mathematics in the United States of America 1792–1810

Abstract: In the preface to his 1788 textbook, *A New and Complete System of Arithmetic, Composed for the Use of the Citizens of the United States,* Nicolas Pike wrote that "as the United States are now an independent nation, it was judged that a system [of arithmetic] might be calculated more suitable to our meridian than those heretofore published" (p. vii). A glance at the contents of Pike's book reveals that it did have features not usually found in previous school arithmetics—for example, it included sections on logarithms, plain geometry, plain trigonometry, algebra and conic sections, and the sections on vulgar and decimal fractions appeared early in the book. But, Pike's book was rejected by many teachers as too difficult for most school children, and despite repeated calls for authors to write arithmetic textbooks which were specially designed for North American students, the most popular of the published arithmetics continued to emphasize the *abbaco*, commercially-oriented, arithmetic curriculum inherited from Europe. This chapter pays attention to immediate effects, during the period 1792–1810, that the introduction of a Federal, decimalized currency had on the author-intended arithmetic curriculum.

Chapter 6: Decimal Fractions in School Arithmetic in Great Britain and North America During the Eighteenth and Nineteenth Centuries

Abstract: This chapter compares the extent to which decimal fractions were part of author-intended and teacher-implemented school mathematics curricula in Great Britain and in North America during the eighteenth and nineteenth centuries. It is assumed that what authors of arithmetic textbooks included in their textbooks was what they hoped, and *intended*, students would study; and that what teachers required their students to write in cyphering books constituted the *implemented* arithmetic curricula for those students. Entries in 472 cyphering books—370 prepared in North America and 102 in Great Britain—were examined and it was concluded that, before 1792, a greater proportion of students in British schools studied decimal fractions than in North American schools. That was an unexpected finding given that most eighteenth-century arithmetic textbooks in both nations included sections on both vulgar and decimal fractions. Analysis also revealed that, in the nineteenth century, the proportion of North American students who studied decimal fractions at schools increased but in Great British it did not. The different levels of emphasis on decimal fractions in the two nations are explained through the theoretical lens of lag time.

Chapter 7: Decimal Fractions and Curriculum Change in School Arithmetic in North America in the Eighteenth and Nineteenth Centuries

Abstract: The main purpose of this final chapter is to answer each of the five research questions identified in Chapter 1. In the eighteenth century the British government had transplanted its sterling pounds, shillings, pence and farthings firmly into the British colonies in North America and since the *abbaco* curriculum followed by those who taught school arithmetic was deeply concerned with money calculations, implemented curricula for school arithmetic in North America asked students to perform countless money calculations based on sterling currency. All of that changed after Thomas Jefferson's (1784) successful efforts to introduce a decimalized currency as Federal money, for then the new nation's schools, and teachers were called upon to re-focus their curricula so that new forms of money calculations would be included. However, the new focus was blurred by Congress's decision not to introduce a Federal decimalized system of weights and measures, and the decision to allow the states to continue to use their own versions of sterling money as "legal" currency. As a result, the shift in emphasis resulting from the courageous decision to introduce decimal currency was slow to take hold. The chapter concludes with commentary on the limitations of the study, and with the identification of researchable questions that scholars might profitably address in the future.

Preface

In 1970 the National Council of Teachers of Mathematics titled its 32nd yearbook *A History of Mathematics Education in the United States and Canada.* Both of the editors of that yearbook—Phillip S. Jones and Arthur F. Coxford—were University of Michigan mathematics professors who had well-documented scholarly interests in mathematics education, including the history of mathematics education.

Jones and Coxford (1970) chose 1821 as the final year of what they identified as the first main period in the history of mathematics education in America. They chose 1821 as the end-point because that was the year Warren Colburn, published his *Arithmetic on the Plan of Pestalozzi with some Improvements* (Colburn, 1821). Colburn's famous text was aimed at teachers of children aged from about 5 or 6 years and, according to Jones and Coxford, its publication marked the first time when mathematics education in America showed concern for pedagogical matters. Jones and Coxford argued that before 1821 the dominant approach to pre-college mathematics education had been through the use of cyphering books in which individual students merely copied and memorized rules, and solved exercises. Jones and Coxford (1970) pointed out that 200 years ago students studying mathematics in American schools rarely owned textbooks—although some of their teachers did own textbooks.

Aside from their unfortunate error of labeling Nathan Daboll, a famous American citizen whose arithmetic textbook was extremely popular between 1800 and 1850, as being English (Jones & Coxford, 1970, p. 45), the 1970 NCTM *Yearbook* did a reasonable job of describing and interpreting mathematics education developments in the United States of America during the nineteenth century. We now know, though, that there was much more to the cyphering tradition than what Jones and Coxford told us in 1970 (see Ellerton & Clements, 2012, 2014).

The subtitle of this present book indicates that the years between 1775 and 1810 have been "neglected" so far as the history of school mathematics in North America is concerned. The subtitle was chosen in order to convey the proposition that something really important happened with respect to mathematics education in North America more than a decade before Colburn's (1821) book was published—and this book investigates that "something."

The "something" we are talking about was the stimulus given to the study of decimal fractions in schools by the introduction of decimal currency as "Federal money" around 1790. Mathematics education research is often characterized as being primarily concerned with investigations into the "teaching and learning of mathematics," but it should be obvious that the quality of mathematics education in a school also depends on the mathematics that students are expected to learn—in other words, the *implemented mathematics curriculum* in that school. In this book we argue, from an analysis of a large data set, that the fillip given to the study of decimal fractions brought about by the introduction of decimal currency *caused* an important change to occur in the type of mathematics that students were asked to learn. We further argue that the architect of the plan to introduce a Federal form of decimal currency, Thomas Jefferson, knew that a decision to mandate decimal currency as the Federal currency would fundamentally change mathematics curricula in schools.

Readers of this Preface may be beginning to think that we are overemphasizing the importance of the impetus given to decimal fractions by the introduction of dollars and cents. To such a point of view we would reply that from our perspective the single most important development in the history of school mathematics was the creation of the Hindu-Arabic numeration system (including zero), and that the mathematical extension of that system toward the end of the sixteenth century to cover decimal fractions proved to be of great importance to research mathematics, service mathematics, and to school mathematics.

In the seventeenth and eighteenth centuries most authors of school arithmetic textbooks included chapters on decimal fractions—and since arithmetic textbooks written by British authors dominated the school textbook market in the British North American colonies, it has been commonly thought that decimal fractions were widely taught in North America in the seventeenth and eighteenth centuries. In this book we show, by analyzing cyphering-book entries, that before 1790, students in North American schools rarely studied decimal fractions. In other words, the *textbook-author-intended* arithmetic curriculum at the time did not match the *teacher-implemented* mathematics curriculum. That state of affairs gradually changed after the introduction of decimalized Federal money. In Great Britain, however, where no decimal currency was introduced, cyphering-book evidence indicates that even well into the nineteenth century decimal fractions were not much studied in the schools.

At the outset we wish to thank Melissa James and Joseph Quatela, of Springer. Without Melissa's and Joe's generous encouragement, support, and promptings, we would not have been able to complete this venture. We also wish to thank librarians and archivists in the Phillips Library (Peabody Essex Museum, Salem, Massachusetts), the Butler Library (Columbia University, New York), the Clements Library (University of Michigan), the Houghton Library (Harvard University), the American University (Washington, DC), the Wilson Library (University of North Carolina at Chapel Hill), the Special Collections Research Center in the Swem Library at the College of William and Mary and the Rockefeller Library (both in Williamsburg, Virginia), the New York Public Library, Guildhall Library, London, the London Metropolitan Archives, and the Milner Library (Illinois State University) for locating relevant manuscripts, artifacts, and books for us.

We feel honored that Dr Douglas Wilson, who was founding Director of the International Center for Jefferson Studies at Monticello and is presently Co-Director of the Lincoln Studies Center at Knox College in Galesburg, Illinois, agreed to write the Foreword for this book. We want to thank Dr George Seelinger, the Head of the Mathematics Department at Illinois State University (where we both work) for encouraging us in our research endeavors.

References

Colburn, W (1821). *An arithmetic on the plan of Pestalozzi, with some improvements.* Boston, MA: Cummings and Hilliard.

Ellerton, N. F., & Clements, M. A. (2012). *Rewriting the history of school mathematics in North America 1607–1861.* New York, NY: Springer.

Ellerton, N. F., & Clements, M. A. (2014). *Abraham Lincoln's cyphering book and ten other extraordinary cyphering books.* New York, NY: Springer.

Jones, P. S., & Coxford, A. F. (1970). From discovery to an awaked concern for pedagogy: 1492–1821. In P. S. Jones & A. F. Coxford (Eds.), *A history of mathematics education in the United States and Canada* (pp. 11–23). Washington, DC: NCTM.

Ken Clements and Nerida F. Ellerton
Illinois State University

Chapter 1

1776: Dawn of a New Day in School Mathematics in North America

Abstract: The 1776 Declaration of Independence announced the dawn of a new day, for North America and, indeed, for the world. But did it herald new a new day for school mathematics? After describing how the term "school" was used in 1776 it is pointed out that at that time only a small proportion of North American adults had studied arithmetic beyond the four operations. The chapter examines the views of Thomas Jefferson on school education, in general, and on school mathematics, in particular. It is argued that it was within Thomas Jefferson's genius that a key to transforming school mathematics in the United States of America would be found.

Keywords: 1776; Decimal currency; Declaration of Independence; Education of slaves; Girls and mathematics; History of school mathematics; Metric system; Thomas Jefferson

A New Day

We hold these truths to be self-evident, that all men are created equal, that they are endowed by their Creator with certain unalienable Rights, that among these are Life, Liberty and the pursuit of Happiness—That to secure these rights, Governments are instituted among Men, deriving their just powers from the consent of the governed—That whenever any Form of Government becomes destructive of these ends, it is the Right of the People to alter or to abolish it, and to institute new Government, laying its foundation on such principles and organizing its powers in such form, as to them shall seem most likely to effect their Safety and Happiness.

With these words, on July 4, 1776, the former colonies became a new nation. The Declaration marked the dawn of a new day for North America and for the world.

The focus of this book is what school mathematics came to look like in this new world. The words of the Declaration of Independence had a mathematical ring about them—the term "self-evident truth" was obviously to be associated with the concept of an "axiom," and with the geometry of Euclid. This was hardly surprising given that Thomas Jefferson, the young man who had been given the responsibility of drafting the Declaration, would, later in his life, write that when he was young, mathematics had been the "passion" of his life (Thomas Jefferson to William Duane, October 12, 1812).

The majesty of the words in the Declaration masked conflicts among those who would sign the Declaration. "All men were created equal," the statement declared, but in 1776 about 20 percent of the 2.5 million people living in the newly conceived nation were slaves, a station in life which had been forced upon them, often through violent means. Many African-background slaves had been seized, in their African homelands, by slave traders, and had been transported, against their wills and usually in appalling sub-human conditions, across the Atlantic Ocean. Many had died on the way. Given those undeniable events of history, any

M. A. (Ken) Clements, & N. F. Ellerton, *Thomas Jefferson and his decimals 1775–1810: Neglected years in the history of U.S. school mathematics*, DOI 10.1007/978-3-319-02505-6_1,
© Springer International Publishing Switzerland 2015

Declaration asserting that all had the unalienable right to liberty and the pursuit of happiness raised challenging questions about how the situation might be forever changed. Yet, even after 1776, George Washington and Thomas Jefferson "owned" many slaves, as did one-third of those who signed the Declaration (Cogliano, 2008; Ellis, 2000; Humphreys, 1991; Wiencek, 2012). The historian is immediately confronted with a problem of "presentism"—however horrendous modern-day readers might view the treatment of slaves in colonial times, how legitimate is it to judge past actions, and people, in terms of today's values (O'Brien, 1996; Wilson, 1992, 1996).

Although there are no hard data available on the proportion of adults in the former North American British colonies in 1776 who had formally learned mathematics beyond counting and simple calculations involving the four operations (addition, subtraction, multiplication and division), that proportion was small. Hardly any of the children of slaves went to school—they were not permitted to do so (Gomez, 1998). Most "free" women had not studied arithmetic at school beyond what they had learned in so-called "dame schools." Although about half of the boys began to prepare cyphering books when they were about 10 years old, many of those who "cyphered" learned very little, if anything, beyond the four operations of arithmetic. Often their teachers knew very little mathematics. In New England colonies, especially, most of the boys who attended school did so in winter months only—for at other times their labor was needed to help their families survive. In selective schools, like the Boston Latin School, which boys attended all year round, most available school time was dedicated to the Classics or to religion (Ellerton & Clements, 2012, 2014).

To what extent, and how quickly, did all this change after 1775?

Contentious Issues

This book is not specifically about the Declaration of Independence—rather, the main focus will be on school mathematics in the United States between 1775 and 1810. Nevertheless, it is important to recognize that the Declaration of Independence announced that, whatever the events of the past, the Founding Fathers thought of the Declaration as ushering in a new day.

Historical perspective makes it clear that those given the responsibility for defining the principles for the new era were not given a blank slate on which they could write anything that they wanted to write. Among the most challenging and potentially divisive issues that they were expected to resolve were:

- How should land and sea forces capable of defending the new nation be formed?
- How could 1500 miles of Atlantic coastline be fortified?
- How could enough income be generated to finance everything that was needed to create a new nation?
- Was a national currency needed?
- Was a national policy defining weights and measures needed?
- Could friendly relationships be established with indigenous Indian nations, whose territories were much desired?
- Could friendly relationships be established with France, and if so, under what conditions?
- Could, and should, friendly relationships be re-established with Great Britain?
- Would slavery be condoned within the new nation?

- Was a national system of schools, with a nationally-prescribed curriculum, desirable? Or, would education policy, and curricular decisions, be decided by each of the separate states, or locally, within individual states?

There were many other contentious issues, but decisions reached with respect to three of those in the above list were likely to affect profoundly the future of mathematics education in the United States of America. The most obvious in that respect was the last, which was concerned with whether a national system of schools, with a nationally-prescribed curriculum, was desirable. Although less obvious at the time, the other two issues that would prove to be highly relevant to the future of mathematics education were the issues relating to national coinage, and to weights and measures. Interestingly, one of the Founding Fathers—Thomas Jefferson—would be a key player in considerations relating to each of the three next five chapters in this book (Chapters 2 through 6) will draw attention to Jefferson's contributions to the coinage and weights and measures issues, and to how the decisions on these issues profoundly affected school mathematics in the United States of America.

But, before discussing attempts to improve school mathematics it will be helpful to reflect on what was embraced by the term "school," and what schools were like in the British colonies in North America in the seventeenth and eighteenth centuries, particularly with respect to the teaching and learning of arithmetic.

What Was a School?

Schooling During the Eighteenth Century in the British North American Colonies

The concept of "school" in North America during the seventeenth and eighteenth centuries included within its ambit "academies," "apprenticeship schools," "common schools," "dame schools," "evening schools," "grammar schools," "local schools," "private schools," "public schools," "subscription schools," and "writing schools" (Cremin, 1970), as well as more specialized establishments like "dance schools," "elocution schools," and "navigation schools." In this book a narrower interpretation of the word "school" than what is implied by that collection of terms will be adopted—so that any formal education environment in which at least one "teacher" regularly met with at least one "student," at an agreed place, for the purpose of helping the student(s) to learn facts, concepts, and skills, from at least one of reading, writing, or arithmetic, will be regarded as having been a school (Ellerton & Clements, 2012). This definition implies that schools did not need to offer formal tuition in any form of mathematics. Higher-level colleges—such as Columbia, Harvard, William and Mary, and Yale—will not be regarded as "schools," for during the eighteenth and nineteenth centuries such higher-level institutions were usually sharply distinguished from "schools."

School mathematics in the European colonial settlements of North America during the period 1607–1776 operated at three main levels. At the lowest level, in so-called "dame schools," young children became acquainted with biblical stories, and learned to say the alphabet, to recognize letters, and to count (Harper, 2010; Perlmann & Margo, 2001; Putnam, 1885; Ryan & Cooper, 2010). At the intermediate level—at a writing or local subscription school, for example—students continued to learn to read and some learned to write and to "cypher." Girls who attended these schools learned to sew and some, usually from well-to-do families, were taught polite accomplishments, fine needlework, manners, and literature

(A Lady of Massachusetts, 1798; Burton, 1794: Lutz, 1929). In schools at the highest level, such as at a Latin school, some students cyphered "beyond the rule of three" (Ellerton & Clements, 2012). From about the age of 10, some boys and a much smaller proportion of girls prepared cyphering books in which they recorded factual information and used the four operations on whole numbers to solve problems (Ellerton & Clements, 2012). Only a small proportion of youth—less than 10 percent—ever attended schools at this highest level.

Whereas the dame schools were attended almost exclusively by children aged between 3 and 8 years, schools at the two higher levels accepted children up to about 18 years. In the New England colonies and in New York and Philadelphia, boys from well-to-do families who attended so-called "Latin grammar" schools prepared for college. They studied Latin, the early histories of Greek and Rome, religious knowledge, and, sometimes, arithmetic (Morison, 1956). In the north-eastern colonies the attendance of boys was usually confined to winter months (Cremin, 1970), but in the South, the extensive use of slave labor meant that plantation owners' children were often freed from working on menial tasks, and were able to attend schools or to receive private tuition, from governesses or "preceptors," throughout the year (A Lady of Massachusetts, 1798; Cremin, 1970).

Arithmetic in dame schools, and the hornbook. Forms of arithmetic taught in the dame schools rarely went beyond familiarizing children with the Hindu-Arabic numerals—specifically 1, 2, 3, 4, 5, 6, 7, 8, 9 and 0—the emphasis being on helping the children to learn to count and to recognize and, perhaps, write the numerals. An intermediate, local writing school, if it existed, had the task of making sure children could read, write and calculate with small numbers. It was only in larger towns that higher-level writing schools, or grammar schools, or both, were to be found. In writing schools, most boys, and probably about one-fifth of the girls, who were at least 10 years of age, prepared handwritten cyphering books. Most of the girls did not prepare cyphering books—it was customary for girls to prepare sewing samplers and to learn fancy embroidery techniques (Cremin, 1970; Earle, 1899; Edmonds, 1991; Ellerton & Clements, 2012; Monaghan, 2007; Ring, 1993; Swan, 1977). If a grammar school existed, it had the task of preparing boys in the classics, especially in Latin, in order that there would be a steady flow of young men eligible to attend college and qualify to become lawyers, or physicians, or, especially, clergymen.

At a dame school the small sons and daughters of European-background settlers—but rarely indigenous Americans, or indentured servants, or slaves—would daily be placed in the care of a "dame" (often a single woman, or a widow). For a small fee, the dame received children at her home, where she was expected to maintain discipline and help the children learn to read and recite the Lord's Prayer. It was understood that provided she carried out these supervision/teaching duties effectively she should feel free, while attending to the children in her care, to complete household chores such as cleaning, sewing and cooking. Often the children would be expected to help with the chores. Each day the dame would read to the children from a Bible and, perhaps, from a primer. Older children—those seven or eight years—might take turns at reading also, and all children would be expected regularly to recite the alphabet. Some would count and answer verbally-presented "sums" (Earle, 1899; Monaghan, 2007). Paper was scarce and expensive, so writing materials were not usually available, but many children brought hornbooks to school. In well-to-do societies a wider range of learning aids was sometimes available in the dame schools.

An image of a dame school in an affluent neighborhood in Great Britain, reproduced from Tuer (1896, p. 115), is shown in Figure 1.1. Note the children's shoes, the quality of their dress, the elaborate furniture, and the naughty boy with a dunce's hat serving time standing on a chair. In less affluent neighborhoods, in Great Britain and in North America, many schoolchildren did not wear shoes to school, even in winter months. The women who conducted dame schools, and the private tutors and governesses of young children employed by well-to-do families, were expected, among other things, to lay the foundation for the development of number sense.

Figure 1.1. Image of an early dame school (from Tuer, 1896, p. 115).

Hornbooks were key learning aids in dame schools. A hornbook was a smallish paddle, often made of wood, with a single printed page on one of its rectangular sides (see Figure 1.2). This page usually ranged in size from about 4 inches by 3 inches to about 7 inches by 5 inches, and the side with the writing was protected by a translucent piece of horn held by tacks on its edges. A hornbook would often be attached to student's waist by string (as was the case with the girl on the left side in Figure 1.1). Typically, a hornbook would show the alphabet, a list of the vowels, and the Lord's Prayer. It might also include a list of the Hindu-Arabic numerals. The hornbook shown in Figure 1.2a, b, which was probably handmade in North America—possibly around 1700—was unusual because it included the Roman numeral equivalents of the Hindu-Arabic numerals as well as the Hindu-Arabic numerals (Miter, 1896; Tuer, 1896).

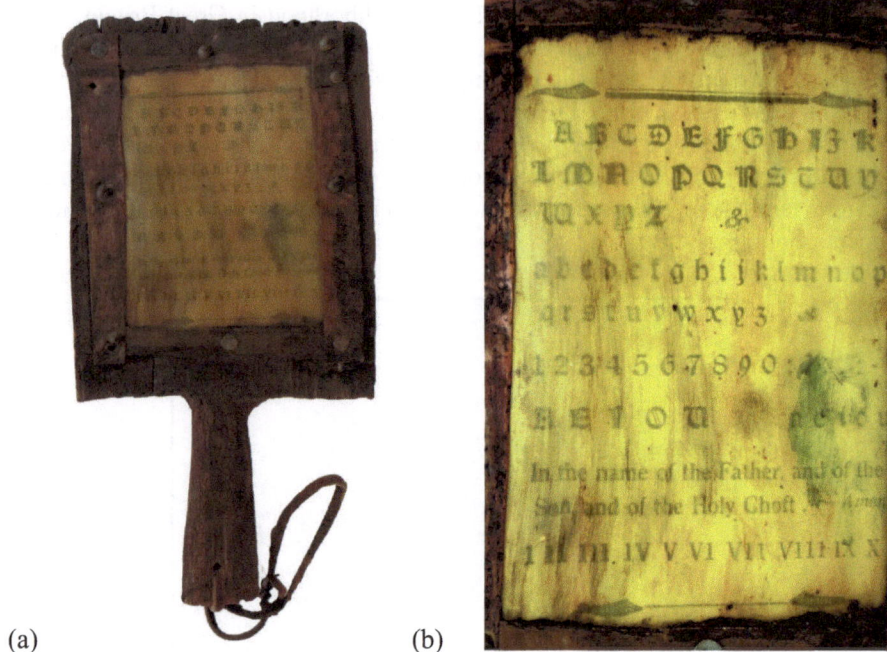

(a)   (b)

Figure 1.2. A hornbook (probably constructed and used in the late-seventeenth or in the eighteenth century in North America).

The authors own the hornbook shown in Figure 1.2a, b.[1] It is, perhaps, the only extant hornbook that was constructed and used in North America during the colonial period (Plimpton, 1916; Tuer, 1896). Usually hornbooks included a cross at the left of the first row (before the "A"), and the first row was accordingly called the "criss-cross" row—but the hornbook shown in Figure 1.2 does not have such a cross. Many hornbooks did not show numerals (Tuer, 1896), and the extent to which hornbooks were used to help young children develop *number sense* has not been studied. It *is* known, though, that young children who proceeded to higher levels of education were typically expected not only to be able to read and write numerals, but also to be able to count and calculate (Ellerton & Clements, 2012).

Local preparatory schools and other schools where arithmetic was part of the curriculum. Many of the early colonists, especially those in New England, wanted their children to learn to read, write and cypher. Some colonists were also keen that a sufficient number of "capable" students would study the classics, especially Latin, to ensure that there would be a guaranteed and steady source of literate clergymen (Cremin, 1970). This latter

---

[1] Expert examination of the hornbook shown in Figure 1.2 led to the conclusion that it is "very old" and was constructed from American white pine. The last owner of the hornbook said that it had previously been owned by a woman who, after serving as a missionary/teacher to indigenous Americans for many years, was given the hornbook as a parting gift. The hornbook could have been used by early missionaries in the seventeenth or eighteenth centuries. According to George Littlefield (1904), often Puritan missionaries, such as John Eliot (see Cogley, 1999), deliberately omitted the cross when constructing hornbooks, for they did not want to expose Native American children to "forms of idolatry."

perceived need resulted in the early establishment in 1635 of the Boston Latin School and of Harvard College, and, in 1660, of the Hopkins School, in New Haven, Connecticut (Cremin, 1970). But, those in control of grammar schools in which Latin was a key curriculum component often regarded their schools as being for children of the élite, and little attention, if any, was given to arithmetic or to any other branch of mathematics (Seybolt, 1935).

Although in the mercantile and maritime cities of Boston, New York, Philadelphia, Salem (Massachusetts), Baltimore, Newport and Providence, it made some sense for children to want to become business clerks, or reckoners, or navigators, or surveyors, most parents thought that survival depended on their getting as much assistance from their children as possible in their daily family-related activities. Furthermore, those who wanted their children to proceed to higher studies had to weigh that desire up against likely costs. Even if qualified tutors could be found, payments for tuition in cyphering were likely to be prohibitive (Ellerton & Clements, 2012). That was especially true of families living some distance from sizeable towns.

According to Money (1993), by 1740 there were about 50 private tutors within the European settlements in North America who advertised that they were willing to receive students for cyphering. Some of the names of those tutors are listed in Karpinski (1980) and Seybolt (1935). The actual number of persons offering classes was likely to have been many times the numbers who advertised in colonial newspapers for students. Indeed, it is likely that there were hundreds of itinerant teachers moving from village to village, offering tuition in reading, writing and occasionally cyphering. Some of these itinerants did not read, write, or cypher well, themselves. And, for most children it was difficult to learn arithmetic without the aid of a tutor because textbooks were not readily available. Even if textbooks were available, the complex language that authors used to describe rules and cases rendered them incomprehensible for most unassisted readers—and that statement applied as much to inexperienced teachers, who had studied very little arithmetic in the past, as it did to school students.

In the eighteenth century there were far more advertisements in major North American newspapers offering tuition in "arithmetick" (as it was usually written) than there were advertisements for printed arithmetics (Monaghan, 2007). At that time, arithmetical education was based on students preparing cyphering books. Historians like to mention well-known authors of early European arithmetics (e.g., Robert Recorde, Edmund Wingate, Edward Cocker, James Hodder, William Leybourn, Pieter Venema, and George Fisher), but it is likely that only a small proportion of students who prepared cyphering books owned, or had ready access to, printed arithmetics (Goodrich, 1857).

Many of the students who studied arithmetic were apprentices. During the seventeenth and eighteenth centuries in the European colonies on the east coast of North America, large numbers of male apprentices learned to read, write and cypher in evening classes (Bremner, 1971; Douglas, 1921; Seybolt, 1917, 1921). Statements in legally enforceable indentures often required masters to ensure that their apprentices learn to read, write and cypher (Seybolt, 1917). Instruction could be given by masters themselves but, if they were unable to do this then apprentices would be expected to attend day or evening classes.

Other aspects of mathematics studied in schools. Branches of mathematics other than arithmetic—such as algebra and geometry—were rarely studied in local schools. In some of the specialized schools for prospective navigators or surveyors elementary forms of geometry—especially geometrical constructions based on straight edges (rulers) and compasses—were studied, but in the common schools arithmetic was the only aspect of mathematics that was dealt with on a regular basis (Ellerton & Clements, 2012, 2014; Jackson, 1906; Kilpatrick & Izsák, 2008; Monroe, 1917; Simons, 1924, 1936; Sinclair, 2008).

Definition of the term "cyphering book." In this book the term "cyphering book" is used to refer to handwritten manuscripts which focused on mathematical content and had all of the following properties:

1. Either the manuscript was written by a student who, through the act of preparing it, was expected to learn and be able to apply whatever content was under consideration; or, it was prepared by a teacher who wished to use it as a model which could be followed by students preparing their own cyphering books.
2. Usually, all entries in the manuscript appeared in ink—as handwritten notes, or problem solutions, or as illustrations. Headings and sub-headings were presented in decorative, calligraphic style. Occasionally, water-color illustrations were included.
3. The manuscript was dedicated to setting out rules, cases, model examples and exercises associated with a sequence of mathematical topics. Although most cyphering books were specifically concerned with arithmetic, especially commercial arithmetic, a few were dedicated to algebra, or geometry, or trigonometry, or to mathematics associated with mensuration, navigation, surveying, fortification, etc.
4. The topics covered were sequenced so that they became progressively more difficult. The content also reflected the expectation that, normally, no child less than 10 years of age would be assigned the task of preparing a cyphering book (Ellerton & Clements, 2014, p. 1).

Sometimes the spelling "ciphering" was used in preference to "cyphering," and that was particularly the case in the nineteenth century. Occasionally the term "copybook" was used to denote a cyphering book, but "copybook" was a more general term used to describe any book in which the writing was by hand and which had been copied entirely from some other text.

Thomas Jefferson's Bills for the More General Diffusion of Knowledge in Virginia

Jefferson on School Education in Virginia

Between 1776 and 1784, Thomas Jefferson and other leading Virginians set their minds to effecting far-reaching revisions of state law relating to school education. As a young wartime Governor of Virginia between 1779 and 1781, Jefferson drafted bills for a "More General Diffusion of Knowledge." He made it clear that he believed that Virginia should establish a system of education which would foster state-supported education at three levels. At the lowest level, in the state schools, most of the day-by-day decisions were to be placed in the hands of local committee members, parents, and teachers. Although a skeleton, "core" curriculum might prevail, the day-by-day curricula in the schools were to be decided locally. But, despite the fact that education bills similar to those drawn up by Jefferson were placed before the Virginia legislature on several occasions, they were never passed (Boyd, 1950b).

In his proposed legislation, Jefferson called for the establishment of a state education system which featured three distinct grades of education:

1. Elementary schools, at which instruction would be available, *gratis*, for all "free" children;

2. Academies, where a middle level of instruction, calculated for the common purposes of life, would be offered; and
3. Colleges, whose curriculum would include the sciences generally, in their highest degree (Boyd, 1950b; Conant, 1962).

Jefferson envisaged a stratified system of state-controlled education which divided the counties into wards, or "hundreds," with "each ward functioning as a 'little republic' in which the citizens would provide for an elementary school to which 'all the free children,' male and female, would be admitted without charge" (Urban & Wagoner, 1996, p. 72). Although the first-level, elementary, schools were to be made available to "free" children—but not to the children of slaves—parents would not be compelled by the state to send their children to an elementary school, even though instruction at such a school would be free. So far as school mathematics was concerned, no thought was given to the possibility of teaching any branch of mathematics other than arithmetic.

The second-level "grammar schools" or "academies" would have boarding-school facilities. They would be created in every county within the State, and would be subsidized by the State. Proprietors would be required to accept, without charge, the most outstanding boy from each of the first-level schools in the county, and children whose parents were prepared to pay for tuition would also be permitted to send their child—male or female—to such a school. Scholarship students in these academies would be culled during the first two years of what would be a six-year program. According to Jefferson (1784), by this means "twenty of the best geniuses" would be "raked from the rubbish annually, and be instructed, at the public expense, so far as the grammar schools go" (quoted in Peden, 1955, p. 146). Scholarship students who survived this culling process would be entitled to remain at the school for six years, all the time receiving full financial support; then half of the graduates of the grammar schools would be given state scholarships to the College of William and Mary at Williamsburg (Urban & Wagoner, 1996).

There has been much analysis and commentary on what Jefferson hoped to achieve with these bills—even though they were never passed (see, e.g., Boyd, 1950b; Conant, 1962; Honeywell, 1931; Urban & Wagoner, 1996). It would not be profitable to comment further on the bills here except to note, briefly, three aspects which relate to one of the main themes of this present work—specifically, Jefferson's contributions to the subsequent development of mathematics education in the United States of America.

The first noteworthy aspect is that although the Declaration of Independence declared that it was self-evident that "all men are created equal, that they are endowed by their Creator with certain unalienable Rights, that among these are Life, Liberty and the pursuit of Happiness," Jefferson did not believe that that axiomatic stance applied to educational opportunity so far as girls and the children of slaves were concerned. Jefferson's scholarship proposals allowed all girls to attend schools for three years at state expense, but beyond that scholarships would be available to males only.

So far as children of slaves were concerned, Jefferson never introduced legislation by which they would be entitled to receive scholarships at any level of education. In 1791, Benjamin Banneker, a free Black trained as a mathematician, clockmaker, and surveyor, sent Jefferson a copy of his *Almanac* in an effort to change Jefferson's views on Blacks' intellectual capacities. In the accompanying letter, Banneker pleaded with Jefferson to live up to the ideals of the Declaration of Independence. Jefferson never provided, at least directly, the assistance that Banneker requested (see *Benjamin Bannaker's New-Jersey, Pennsylvania,*

Delaware, Maryland and Virginia Almanac, or Ephemeris, for the Year of our Lord 1795.
Wilmington, Delaware, 1795, held in the Rare Book and Special Collections Division of the
Library of Congress). It is true that in a 1796 letter to Robert Pleasants, a Quaker, member of
the Virginia legislature, Jefferson asked Pleasants to propose a Bill that would permit "slaves
destined to be free" to be eligible for free instruction. However, although the measure was
proposed as an amendment to a bill, the amendment was removed from the final form of the
bill. In a handwritten letter to Nathaniel Burwell, dated March 14, 1818 (held in the
Manuscript Division of the Library of Congress), Jefferson wrote that for his own daughters
he thought it "essential to give them a solid education which might enable them, when they
become mothers, to educate their own daughters."

Virginians had good reasons to doubt whether a centralized system of education, with
standard curricula, was appropriate for their State. Although Virginia was the most populous
State in the new Union, it had no large cities. Data generated by the first U.S. Census, in
1790 indicated that across the 13 states only 1 person in every 20 lived in an urban area with
a population which was more than 2600 (U.S. Department of Commerce, Bureau of the
Census, 1975). Virginia, with a population of about three-fourths of a million "free" people
and about one-fourth of a million slaves, had easily the largest population of all 13 states, and
by far the greatest number of slaves. Yet, the largest town in the State was Richmond, with a
population of less than 4000. Clearly, the people were scattered across the state on
plantations or in small communities. The kind of school education system the State might
like to support would have been very different from the kind of school system that
Massachusetts, say, would have liked. Massachusetts was much more heavily settled, had
four urban centers larger than Richmond (Boston, Salem, Marblehead and Newburyport), a
well-established college (Harvard College, in Cambridge, Boston), thriving shipping ports at
Boston and Salem, and relatively few slaves to lessen the labor load on farming communities
(Ellerton & Clements, 2012).

But, Jefferson did give careful thought to two other matters that, potentially at least,
were likely to have an impact on school mathematics. Those two matters were the
establishment of a national system of currency, and the establishment of a national system of
weights and measures.

Before moving to those important considerations, however, it will be useful to reflect
that there were some North Americans who did see the need for nationally centralized
education policies. One such person was Noah Webster, whose efforts to standardize
American spelling and pronunciation would prove to be largely successful, despite the fact
that they were not supported by national legislation.

Who Should Control the Schools?

Although much has been written about events which culminated in the U.S.
Constitution of 1787 (see, e.g., Bowen, 1966), the most important point in relation to the
main themes to be developed in this book is that that Constitution made no direct reference to
school education (Good, 1960). This was undoubtedly because each state wanted to control
the forms of education that were offered in schools within its own borders. Kamens and
Benavot (1991) were wrong when they claimed that mathematics became a compulsory
secondary school subject in the United States in 1730. Aside from the fact that, obviously,
the United States did not exist as a formal entity at that time, there were never any formal

agreements between the British colonies, or between the colonies and the British government, on curricular matters. In any case, the term "secondary school" was not well defined.

Kamens and Benavot (1991) were also wrong when they claimed that in 1790 arithmetic was a compulsory subject in common schools located within the United States. From the 1630s onwards, Massachusetts demanded that elementary public schools should be available in which young children could learn to read, write, and cypher (Cremin, 1970). But in the colonial periods and in the early Federal period none of the states thought that there needed to be a national system of education, with mandatory national curricula. In 1776, some states wanted to leave the curricula, and conduct of schools to local communities; others were prepared to allow a measure of state control over curricula; but no state wanted the new national political entity to have control over schools or their curricula (Cremin, 1970; Cubberley, 1962). In fact, during the period 1776–1790 each of the 13 states moved to define, confirm, and to extend different levels of state government control over its own schools (Good, 1960). All 13 states insisted that matters such as who should be required to attend school, or who should define curricula, or who should be allowed to teach in the schools, should be decided within each individual state (Urban & Wagoner, 1996).

An interesting aspect of Jefferson's planned arithmetic curriculum for the second level of his three-level school curriculum was that all the students would be expected to study vulgar fractions and decimal fractions, as well as square and cube roots. Although in principle, the control of state-supported schools, and of curricula, were to be left in the hands of locals, that principle did not, at least in Jefferson's mind, prevent the state from specifying certain core knowledge which should be studied. A twenty-first-century reader would be unlikely to find anything unusual in that—but, analyses provided in this book will show that, for example, in 1779 only a tiny proportion of North American school students formally studied either vulgar or decimal fractions, both of which were to be part of Jefferson's core curriculum for arithmetic.

Noah Webster's Call for a National Agenda for Education

There were some scholars who favored a move towards centralizing the control of curricula. Thus, for example, Noah Webster (1789), in his *Dissertations on the English Language*, urged the new nation to move toward standardizing its language usage, including the spelling and pronunciation of words. Webster (1789) wrote:

> The United States were settled by emigrants from different parts of Europe. But their descendants mostly speak the same tongue; and the intercourse among the learned of the different States, which the Revolution has begun, and an American Court will perpetuate, must gradually destroy the differences of dialect which our ancestors brought from their native countries. This approximation of dialects will be certain; but without the operation of other causes than an intercourse at Court, it will be slow and partial. The body of the people, governed by habit, will still retain their respective peculiarities of speaking; and for want of schools and proper books, fall into many inaccuracies, which, incorporating with the language of the state where they live, may imperceptibly corrupt the national language. Nothing but the establishment of schools and some uniformity in the use of books, can annihilate differences in speaking and preserve the purity of the American tongue. A sameness of pronunciation is of considerable consequence in a political view; for provincial accents are disagreeable to strangers and

sometimes have an unhappy effect upon the social affections. All men have local attachments, which lead them to believe their own practice to be the least exceptionable. Pride and prejudice incline men to treat the practice of their neighbors with some degree of contempt. Thus small differences in pronunciation at first excite ridicule—a habit of laughing at the singularities of strangers is followed by disrespect—and without respect friendship is a name, and social intercourse a mere ceremony.

These remarks hold equally true, with respect to individuals, to small societies and to large communities. Small causes, such as a nick-name, or a vulgar tone in speaking, have actually created a dissocial spirit between the inhabitants of the different states, which is often discoverable in private business and public deliberations. Our political harmony is therefore concerned in a uniformity of language.

As an independent nation, our honor requires us to have a system of our own, in language as well as government. Great Britain, whose children we are, and whose language we speak, should no longer be our standard; for the taste of her writers is already corrupted, and her language on the decline. But if it were not so, she is at too great a distance to be our model, and to instruct us in the principles of our own tongue. (p. 20)

As a result of his untiring efforts, Webster was able to achieve much with respect to his desire to standardize "American" spelling and pronunciation. Webster was instrumental in not only changing the ways Americans pronounced and spelt words—he also induced large changes in school curricula, in all states (Ellerton & Clements, 2008).

It will be important, in this book, to investigate whether, in the early Federal period, there was a similar concerted attempt to standardize school mathematics within and across the 13 states.

Moves Toward Decimalization of Currency and Weights and Measures

Preliminary Moves Toward a New Currency

The Continental Congress was established by the former British colonies in North America in 1774, and during the first two years of the American Revolution it became the formal means by which the American colonial governments coordinated their resistance to British rule. Given war-time financial demands, it was important that some control over currency be established, and in April 1776 a committee of seven was appointed by Congress "to examine and ascertain the value of several species of gold and silver coins, current in these colonies, and the proportion they might bear to Spanish milled dollars" (Boyd, 1950a, p. 518). This report, which was prepared largely by George Wythe, Roger Sherman and James Duane, was tabled on May 22, 1776, but no action was taken until July 24, 1776 (Boyd, 1953), when the acceptance of the Declaration of Independence, 20 days earlier, made it imperative that a currency system for the new independent nation be created. At the meeting of the Continental Congress on July 24 it was agreed that "the report of the committee on gold and silver coins be recommitted" and that "Mr Jefferson be added to the said committee (Boyd, 1950a, p. 518*n*). The revised report was duly prepared and placed before Congress on September 2, 1776.

According to Julian Boyd (1953), Thomas Jefferson influenced greatly the content, the wording and the form of the revised report on coinage. Whereas the original report, prepared by Wythe, Sherman and Duane, had expressed values of coins of different nations in vulgar fractions of the Spanish dollar, the new report gave the values in "decimal notation in dollars and parts of a dollar" (Boyd, 1950a, p. 516). According to Boyd (1953), "this evidently, was the first effort to employ decimal reckoning in the money system of the United States" (p. 152). Thus, for example, whereas the original report had used complex common fractions such as $\frac{20}{139}$ and $\frac{7404}{15729}$ when making comparisons, Jefferson gave all his comparisons in decimal fractions (which, in the report, were correct to six decimal places—see Boyd, 1950a, p. 517). Jefferson stopped being a member of the Continental Congress on the day the second report was tabled, and it was not until the early 1780s that his interest in coinage issues were officially revived.

Clearly, the first Congress investigation into the values of different gold and silver coins widely used in the colonies provided an introduction to considerations related to the form of currency to be adopted in the new nation. In 1776 each of the 13 states had its own version of sterling currency, with the values of pounds, shillings, pence and quarters (farthings) likely to change when one crossed a border—although the ratios of 4 farthings to a penny, 12 pence to a shillings, and 20 shillings to a pound remained constant.

During the war period (1776–1783) Spanish and Mexican dollars were widely used in the former colonies, and therefore when the question of what form of new currency should be created in the new nation was considered it was only natural that Spanish and Mexican dollars would become an important part of the conversation. The analysis in Chapter 2 will reveal that before the War those who studied arithmetic at school expended much time and effort attempting to comprehend and solve problems involving money, and so any decision on the form of currency was likely to have a huge effect on implemented arithmetic curricula in schools. In 1783 and 1784 Jefferson, who, after a stint as Governor of Virginia once again become a member of Congress, would re-invigorate national interest in the matter. By successfully advocating a decimalized form of currency, he would be responsible for a change that had the potential to influence, profoundly, the content and face of elementary school mathematics in the United States of America.

Preliminary Moves Toward a New System of Weights and Measures for a New Nation

The United States Constitution of 1787 granted Congress the power to "fix the Standard of Weights and Measures." Merely having the power to do so did not, however, make the actual task of "fixing" an easy one. Although the weights and measures used in the former British colonies of North America were mostly English in origin, in actual fact the systems used varied not only from colony to colony, but also from region to region (Bordley, 1789). The variety in the colonies reflected the diversity of customary usage in the European nations from which the colonists had come. The competing units of measure caused confusion in everyday life, and in business transactions.

For example, all of the following capacity measures were used in the colonies: the firkin, kilderkin, strike, hogshead, tierce, pipe, butt, and puncheon. Even when the same name for a unit was used from colony to colony or locality to locality, that unit was not always assigned the same value. A bushel of oats in Connecticut, for example, weighed 28 pounds, but in New Jersey it weighed 32 pounds.

The rationale behind a locality's adoption of a particular measure was often obscure. With the addition of New York to the British colonies in 1664, numerous Dutch measures were introduced. French and Spanish measures were brought into the mix and were used in various parts of North America throughout the seventeenth and eighteenth centuries.

By March 1784, *before* he relocated to Paris as the U.S. Minister Plenipotentiary to negotiate treaties of amity and commerce, Thomas Jefferson had conceived of a plan to decimalize weights and measures in the United States of America. Boyd (1953) located a previously misplaced document (see pp. 173–175)—the original of which was entirely in Jefferson's hand and, although undated was probably written in February or March 1784—showing that even before he went to Paris Jefferson aimed to achieve, in the United States, a decimalized system of weights and measures, and perhaps time. The text of that document, titled "Thoughts on a Decimal Coinage," is reproduced as Appendix *A* to this book. Jefferson clearly wanted to link his decimalized coinage to measures of weights and measures. In "Thoughts on a Decimal Coinage" Jefferson wrote of a "transition from weights to measures" with "rain water weighing a pound, i.e., 10 units, to be put in a cube vessel and one side of that taken for the standard of unit or unit of measure."

If the genesis of the idea of establishing decimalized systems of currency and of weights and measures in the United States came from Jefferson's fertile mind, the first major statement to Congress concerning the need to standardize weights and measures was made by Jefferson's friend, James Madison. In a letter to James Monroe, dated April 31, 1785, Madison wrote that in the regulation of weights and measures it would be highly expedient

> ... to pursue the hint which had been suggested by ingenious and philosophical men, to wit: that the standard of the measure should be fixed by the length of a pendulum vibrating seconds at the equator or any given latitude ... Such a scheme appears to be entirely reducible to practice; and as it is founded on the division of time, which is the same at all times and at all places, and proceeds on other data which are equally so, it would not only secure a perpetual uniformity throughout the United States, but might lead to universal standards in these matters among nations. Next to the inconvenience of speaking different languages, is that of using different and arbitrary weights and measures. (Madison, 1865, pp. 152–153)

It was not long after Jefferson returned to North America, at the beginning of 1790, that the matter of establishing a Federal decimalized system of weights and measures was once again taken up seriously—this time by Jefferson, himself. Jefferson had been appointed Secretary of State by George Washington, and one of his official duties was to draft "a proper plan, or plans, for establishing uniformity in the currency, weights, and measures of the United States (Boyd, 1961, p. 64). It was known that George Washington liked Jefferson's decimal approach to the matter (Boyd, 1961).

The U.S. Congress accepted Jefferson's decimal currency scheme, but rejected his proposed application of decimal fractions to a coordinated system of weights and measures The events which led to, and the politics behind, the rejection of Jefferson's plan for a national system of weights and measures have been well described by Andro Linklater (2003). It suffices here to say that the urgent need to gain a strong and regular cash inflow into national coffers meant that land which was opening up in the West had to be sold quickly, and at the highest possible price. That meant the land had to be surveyed quickly, and the politicians were persuaded that the surveyors would not have been able to cope with any radically new system of land measurement. Chapters 3 and 5 will consider educational consequences of the decimalization of U.S. currency, and Chapters 4, 5, and 6 educational consequences of non-decimalization of weights and measures. Curiously, although the effects of the decisions on currency and weights and measures on school mathematics were large, they have never been systematically studied by historians, from the perspective of school mathematics.

Data generated by studying statements of intended curricula (largely in textbooks) and implemented curricula (largely through cyphering books) will be analyzed in Chapters 2 through 6, and as a result of findings generated by the analyses, conclusions will be reached in the final chapter (Chapter 7).

Research Questions

This main aim of the research described in this book was to answer the following five research questions:

1. To what extent did policies and practices relating to money influence school arithmetic curricula in the North American colonies before 1790?
2. To what extent did policies and practices with respect to weights and measures influence school arithmetic curricula in the United States of America before 1790?
3. To what extent did the place of decimal fractions in U.S. school arithmetic curricula change as a result of the introduction of a decimalized scheme of Federal money around 1790?
4. To what extent did policies and practices with respect to currency and weights and measures play a part in school arithmetic curricula in the United States of America between 1791 and 1860?
5. What are the implications of the findings of this study for mathematics curriculum theory?

The first two of these five questions will be dealt with in Chapters 1 and 2, and the third question in Chapters 3, 5 and 6. Data in relation to Question 4 will be considered in Chapters 4 and 6. Discussion in relation to Question 5 will be offered in Chapter 7, the final chapter. In that chapter answers to each of the research questions will also be given, and implications of the study for mathematics curriculum theory considered. Limitations of the study will also be discussed in Chapter 7, and questions raised for further research.

As far as we know, none of the above questions has been previously addressed by historians. Although our aim throughout the book will be to answer the five questions, we will not be totally constrained by them. Occasionally, we will deviate from the main pathway to consider how our analyses might be of relevance to other important mathematics education issues.

The Lag-Time Theoretical Lens

Figure 1.3, which is from Ellerton and Clements (2014—see p. 323), draws attention to how different groups can interpret mathematics-related problems from different perspectives. Figure 1.3 distinguishes between three intersecting types of mathematics—research mathematics, service mathematics, and mathematics education—and emphasizes that the three forms of mathematics are developed within societies in which "ethnomathematical forces" shape, use, and sometimes modify existing forms of mathematics." During the period

1667–1887, ethnomathematical contexts varied enormously, both within and between nations. Advances in mathematics, and in applications of mathematics, interacting with the needs of evolving communities, changed what Ian Westbury (1980) has referred to as "intended," "implemented," and "attained" school mathematics curricula.

This study will use a lag-time construct to provide a theoretical lens through which data will be examined. Basically, "lag time" is the time that elapses between an event and a subsequent related event. Although a lag-time theoretical lens has not previously been used in the contexts of the history of mathematics education, "lag time" has been made a central concept in relation to theories underlying many other areas of scholarship (e.g., in mathematics and hereditary processes—see Bellman & Danskin, 1954; in economics—see Besomi, 1998; and in chemistry—see Keraliya & Patel, 2014; Wu, 1992). Here, we develop a specific way of using the concept in relation to mathematics curriculum development.

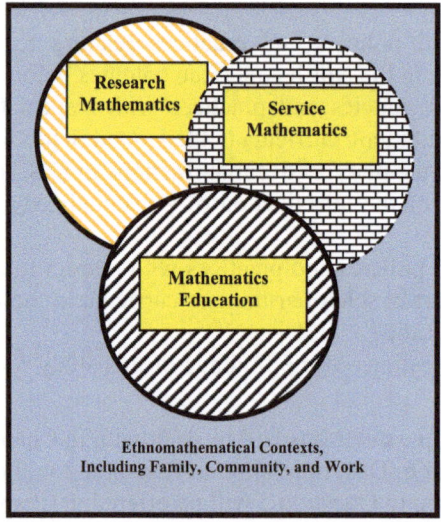

Figure 1.3. Different ways of "seeing" problems or situations that might relate to mathematics (from Ellerton & Clements, 2014, p. 323).

In the case of implemented mathematics curricula, lag time will be defined as the amount of time between when a mathematical development (such as the definition of a decimal fraction, and statements describing how decimal concepts can be applied in mathematics and in society) was first introduced—probably by a mathematician or a practitioner—and accepted by the professional mathematicians, and the time when that development was sufficiently simplified that it was "normally studied" as part of the implemented mathematics curriculum in schools in particular communities.

Lag times for the same set of mathematical concepts (e.g., decimal fractions) will be likely to differ from place to place because the interests, needs, and priorities of "societies-at-large" differ from place to place (Ogburn, 1922). Obviously, lag time will vary within and between communities, depending on a community's willingness, or lack of willingness, to include a concept or principle or skill in the implemented mathematics curriculum of its schools. We recognize that the term "normally studied" in our definition of lag time will likely result in subjectivity, but we hesitate to define the term more precisely. In the case of the present study, which will mainly examine implemented curricula in North America in the eighteenth and early nineteenth centuries, but will also compare curricula in the United States of America and Great Britain between 1667 and 1887, we shall adopt a pragmatic definition by which we assume that the adoption of a development had occurred when that development was formally dealt with in at least 50 percent of the cyphering books prepared within a community.

There is a natural tendency for research mathematicians to believe that they are the only "true" mathematicians, and for "applied mathematicians" to believe that they are the only ones who "have their feet on the ground" doing work which permits mathematics to become useful to society. Likewise, those heavily involved in mathematics education like to believe that they are the ones who have the task of creating curricula and assisting teachers to establish learning environments which will help all students learn the mathematics they will need in their present and future lives. From our perspective, each of the three components of mathematics has a vitally important, but nevertheless different, role to play, and it is counter-productive for one group to think it is somehow more important than the other two.

Some commentators and scholars see any issue related to school mathematics, or apparently related to school mathematics, from a research mathematician's perspective; others see the issue from a service perspective—whether mathematics is being used to explore and solve real-life problems; and others see the issue from the point of view of the teaching and learning of mathematics. Some try to see it from the perspective of the intersections which can be associated with the domains represented in Figure 1.3. Those without much formal mathematical training are likely to try to view problems from other vantage points: they might prefer to tackle problems from the point of view of what could be termed "ethnomathematics." Without necessarily realizing they are engaging in mathematics-related activities, they could use mathematical principles that they have acquired as a result of their participation in everyday life.

Ellerton and Clements (2013) expounded this lag-time theoretical perspective in relation to the development of decimal fractions. Although the earliest theoretical development of decimal fractions concepts is shrouded in the mists of history, it is known that early in the fifteenth century, in Iran, Ghiyath al-Din Jamshid Mas'ud al-Kashi offered a clear description of some of their mathematical properties (Rashed, 1994). According to Boyer (1991), a more complete overview of their *mathematical* significance was offered by the French mathematician François Viète (1579). It was Simon Stevin (1585), the Dutch mathematician/engineer who identified the *practical* significance of decimals: Stevin declared that the introduction of decimal coinage and measures was merely a question of time (De Morgan, 1847; Tabak, 2004).

But the work of research mathematicians had little influence on our thinking as we conducted the research that has generated this book. We were more interested in the influence of those wishing to apply the concept of a decimal fraction in the new United States of America in the years following 1775. It seemed to us, that the words at the bottom of Figure 1.3, "Ethnomathematical Contexts, Including Family, Community, and Work," were particularly important so far as our study was concerned. That is why in this first chapter we have attempted to how political and family circumstances necessarily affected not only what went on in the name of arithmetic in the various kinds of schools in the North American colonies, but also who were permitted, and not permitted, to study arithmetic, or other forms of mathematics. In the mind of Thomas Jefferson, the chief mover in our study, "Service Mathematics" was also important. From Jefferson's perspective the concept of a decimal fraction, and operations on decimals, needed to be brought into service for the community— and to his credit he recognized from the outset that that would have important implications for school mathematics. Pertinent mathematics education issues arising from Jefferson's foresight will be carefully considered throughout this book.

Scholars in the history of mathematics and mathematics education have mainly considered evidence from *author-intended* curricula (as found in textbooks), and in this study such evidence will be taken into account. But, our main sources will be handwritten cyphering books, prepared by students (see, e.g., Ellerton & Clements, 2012, 2014) as well as other primary sources—such as statements found in biographies and novels. That combination of data sources will enable data to be generated, from multiple perspectives, especially with respect to *author-intended,* and *teacher-implemented* curricula. Throughout this book the adjective "author-intended" will often be relaxed to "intended," and, likewise, "teacher-implemented" will be relaxed to "implemented."

References

A Lady of Massachusetts. (1798). *The boarding school, or lessons of a preceptress to her pupils: Consisting of information, instruction, and advice, calculated to improve the manners, and form the character of young ladies.* Boston, MA: I. Thomas and E. T. Andrews.

Bellman, R., & Danskin, J. M. (1954). *A survey of the mathematical theory of time-lag, retarded control, and hereditary processes.* The RAND Corporation, Report R-256.

Besomi, D. (1998) Harrod, and the "time-lag" theories of the cycle. In G. Rampa, S. Stella, & A. P. Thirlwall (Eds.), *Economic dynamics, trade and growth: Essays on Harrodian themes* (pp. 107–143). Basingstoke, UK: Macmillan.

Bordley, J. B. (1789). *On monies, coins, weights and measures.* Philadelphia, PA: Dani

Bowen, C. D. (1966). *Miracle at Philadelphia: The story of the Constitutional Convention May to September 1787.* Boston, MA: Little, Brown and Company.

Boyd, J. P. (1950a). *The papers of Thomas Jefferson* (Vol. 1, 1760–1776). Princeton, NJ: Princeton University Press.

Boyd, J. P. (1950b). *The papers of Thomas Jefferson* (Vol. 2, January 1777 to June 1779). Princeton, NJ: Princeton University Press.

Boyd, J. P. (1953). *The papers of Thomas Jefferson* (Vol. 7, March 1784 to February 1785). Princeton, NJ: Princeton University Press.

Boyd, J. P. (1961). *The papers of Thomas Jefferson* (Vol. 16, November 1789 to July 1790). Princeton, NJ: Princeton University Press.

Boyer, C. B. (1991). *A history of mathematics* (2nd ed.). New York, NY: Wiley.

Bremner, R. H. (1971). *Children and youth in America: A documentary history* (Vol. 1, 1600–1865). London, UK: Oxford University Press.

Burton, J. (1794). *Lectures on female education and manners.* New York, NY: Samuel Campbell.

Cogley, R. W. (1999). *John Eliot's mission to the Indians before King Philip's War.* Cambridge, MA: Harvard University Press.

Cogliano, F. D. (2008). *Thomas Jefferson: Reputation and legacy.* Charlottesville, VA: University of Virginia Press.

Conant, J. B. (1962). *Thomas Jefferson and the development of American public education.* Berkeley, CA: University of California Press.

Cremin, L. A. (1970). *American education: The colonial experience 1607–1783.* New York, NY: Harper & Row.

Cubberley, E. P. (1962). *Public education in the United States.* Boston, MA: Houghton Mifflin Co.

De Morgan, A. (1847). *Arithmetical books, from the invention of printing to the present time.* London, UK: Taylor and Walton.

Douglas, P. H. (1921). *American apprenticeships and industrial education.* PhD dissertation, Columbia University, New York.

Earle, A. M. (1899). *Child-life in colonial days.* New York, NY: The Macmillan Company.

Edmonds, M. J. (1991). *Samplers and samplermakers: An American schoolgirl art, 1700–1850.* New York, NY: Rizzoh.

Ellerton, N. F., & Clements, M. A. (2008). An opportunity lost in the history of school mathematics: Noah Webster and Nicolas Pike. In O. Figueras, J. L. Cortina, S. Alatorrw & A. Mepúlveda (Eds.), *Proceedings of the Joint Meeting of PME 32 and PME-NA XXX* (Vol. 1, pp. 447–454). Morelia, Mexico: Cinvestav-UMSWH.

Ellerton, N. F., & Clements, M. A. (2012). *Rewriting the history of school mathematics in North America 1607–1861. New* York, NY: Springer.

Ellerton, N. F., & Clements, M. A. (2013, September). *The mathematics of decimal fractions and their introduction into British and North America schools.* Paper presented to the Third International Conference on the History of Mathematics Education held at Uppsala University, Sweden.

Ellerton, N. F., & Clements, M. A. (2014). *Abraham Lincoln's cyphering book and ten other extraordinary cyphering books.* New York, NY: Springer.

Ellis, J. R. (2000). *Founding brothers.* New York, NY: Random House.

Gomez, M. A. (1998). *Exchanging our country marks: The transformation of African Identities in the colonial and Antebellum South.* Chapel Hill, NC: University of North Carolina.

Good, H. G. (1960). *A history of Western education.* New York, NY: Macmillan.

Goodrich, S. (1857). *A pictorial history of the United States.* Philadelphia, PA: E. H. Butler & Co.

Harper, E. P. (2010). Dame schools. In T. Hunt, T. Lasley, & C. D. Raisch (Eds.), *Encyclopedia of educational reform and dissent* (pp. 259–260). Thousand Oaks, CA: Sage Publications.

Honeywell, R. J. (1931). *The educational work of Thomas Jefferson.* Cambridge, MA: Harvard University Press.

Humphreys, D. (1991). *Life of General Washington.* Athens, GA: University of Georgia Press.

Jackson, L. L. (1906). *The educational significance of sixteenth century arithmetic from the point of view of the present time.* New York, NY: Teachers College Columbia University.

Jefferson, T. (1784). *Notes.* In W. Peden (Ed.), *Notes on the State of Virginia.* Chapel Hill: University of North Carolina Press for the Institute of Early American History and Culture, Williamsburg, VA, 1955.

Kamens, D. H., & Benavot, A. (1991). Elite knowledge for the masses: The origins and spread of mathematics and science in national curricula. *American Journal of Education, 99*(2), 137–180.

Karpinski, L. C. (1980). *Bibliography of mathematical works printed in America through 1850.* New York, NY: Arno Press.

Keraliya, R. A., & Patel, M. M. (2014). Effect of viscosity of hyprophilic coating, polymer on lag time of atendolol pulsatile press coated tablets. *Journal of Pharmaceutical Chemistry, 1*(1), 1–9.

Kilpatrick, J., & Izsák, A. (2008). A history of algebra in the school curriculum. In C. E. Greenes & R. Rubenstein (Eds.), *Algebra and algebraic thinking in school mathematics: Seventieth yearbook* (pp. 3–18). Reston, VA: National Council of Teachers of Mathematics.

Linklater, A. (2003). *Measuring America: How the United States was shaped by the greatest land sale in history.* New York, NY: Plume.

Littlefield, G. E. (1904). *Early schools and school-books of New England.* Boston, MA: The Club of Odd Volumes.

Lutz, A. (1929). *Emma Willard: Daughter of democracy.* Boston, MA: Houghton Mifflin.

Madison, J. (1865). *Letters and other writings of James Monroe 1769–1793.* Philadelphia, PA: J. B. Lippincott & Co.

Miter (1896, September 8). Our London letter. *The American Stationer, 40*(10), 367–368 and 379.

Monaghan, E. J. (2007). *Learning to read and write in colonial America.* Amherst, MA: University of Massachusetts Press.

Money, J. (1993). Teaching in the market place, or *"Caesar adsum jam forte Pompey aderat":* The retaining of knowledge in provincial England during the 18th century. In J. Brewer & R. Porter (Eds.), *Consumption and the world of goods* (pp. 335–377). London, UK: Routledge.

Monroe, W. S. (1917). *Development of arithmetic as a school subject.* Washington, DC: Government Printing Office.

Morison, S. E. (1956). *The intellectual life of colonial New England.* Ithaca, NY: New York University Press.

O'Brien, C. C. (1996). Thomas Jefferson: Radical and racist. *The Atlantic Monthly, 278*(4), 53–74.

Ogburn, W. F. (1922). *Social changes with respect to culture and original nature.* New York, NY: B. W. Huebsch.

Peden, W. (Ed.). (1955). *Notes on the State of Virginia by Thomas Jefferson.* Chapel Hill, NC: University of North Carolina Press.

Perlmann, J., & Margo, R. (2001). *Women's work? American schoolteachers, 1650–1920.* Chicago, IL: University of Chicago Press.

Plimpton, G. A. (1916). *The hornbook and its use in America.* Worcester, MA: American Antiquarian Society.

Putnam, E. (1885). A Salem dame-school. *The Atlantic Monthly, 55*(327), 53–58.

Rashed, R. (1994). *The development of Arabic mathematics: Between arithmetic and algebra.* Dordrecht, The Netherlands: Kluwer.

Ring, B. (1993). *Girlhood embroidery: American samplers and pictorial needlework, 1650–1850.* New York, NY: Knopf Publishers.

Ryan, K. R., & Cooper, J. M. C. (2010). Colonial origins. In L. Mafrici (Ed.), *Those who can teach* (12th ed.). Wadsworth: Cengage Learning.

Seybolt, R. F. (1917). *Apprenticeship and apprenticeship education in colonial New England and New York*. New York, NY: Teachers College, Columbia University.

Seybolt, R. F. (1921). The evening schools of colonial New York City. In New York State Department of Education (Ed.), *Annual report of the State Department of Education* (pp. 630–652). Albany, NY: State Department of Education.

Seybolt, R. F. (1935). *The private schools of colonial Boston*. Cambridge, MA: Harvard University Press.

Simons, L. G. (1924). *Introduction of algebra into American schools in the 18th century*. Washington, DC: Department of the Interior Bureau of Education.

Simons, L. G. (1936). Short stories in colonial geometry. *Osiris, 1*, 584–605.

Sinclair, N. (2008). *The history of the geometry curriculum in the United States*. Charlotte, NC: Information Age Publishing, Inc.

Stevin, S. (1585). *De Thiende*. Leyden, The Netherlands: The University of Leyden.

Swan, S. B. (1977). *American women and their needlework 1700–1850*. New York, NY: Holt, Rinehart and Winston.

Tabak, J. (2004). *Numbers: Computers, philosophers, and the search for meaning*. New York, NY: Facts on File.

Tuer, A. M. (1896). *History of the horn-book*. London, United Kingdom: Leadenhall Press.

Urban, W. J., & Wagoner, J. L. (1996), *American education: A history*. New York, NY: McGraw-Hill.

U.S. Department of Commerce, Bureau of the Census (1975). *Historical statistics of the United States: Colonial times to 1970*. Washington, DC: Government Printing Office.

Viète, F. (1579). *Canon mathematicus seu ad triangula cum appendicibus*. Paris, France: Jean Mettayer.

Webster, N. (1789). *Dissertations on the English* language: *With notes, historical and critical, to which is added, by way of appendix, an essay on a reformed mode of spelling, with Dr. Franklin's arguments on that subject*. Boston, MA: Isaiah Thomas.

Westbury, I. (1980). Change and stability in the curriculum: An overview of the questions. In H. G. Steiner (Ed.), *Comparative studies of mathematics curricula: Change and stability 1960–1980* (pp. 12–36). Bielefeld, Germany: Institut für Didaktik der Mathematik-Universität Bielefeld.

Wiencek, H. (2012). *Master of the mountain: Thomas Jefferson and his slaves* (New York, NY: Farrar, Strauss and Giroux).

Wilson, D. L. (1992). Thomas Jefferson and the character issue. *The Atlantic Monthly, 270*(3), 37–74.

Wilson, D. L. (1996, October). Counter points. *The Atlantic Monthly* online.

Wu, D. T. (1992). The time lag in nucleation theory. *The Journal of Chemical Physics, 97*, 2644.

Chapter 2
The *Abbaco* Curriculum in Colonial Schools in North America Before 1776

Abstract: This chapter examines the extent to which arithmetic curricula in schools in North America before 1776 were concerned with calculations and problems involving money and weights and measures. Analyses of chapters in arithmetic textbooks (deemed to be "author-intended curricula") and handwritten entries in students' cyphering books (regarded as evidence of "teacher-implemented arithmetic curricula") reveal that tasks related to money and to weights and measures permeated what historians have called the *abbaco* curriculum. That curriculum influenced the way the content of arithmetic was structured and sequenced in education institutions below the college level. The analyses also reveal that although money tasks and weights and measures tasks provided the principal emphases in school arithmetic, many students merely applied rules and methods which they did not understand.

Keywords: *Abbaco* curriculum; Arithmetic in U.S. schools; Ciphering books; Common fractions; Cyphering books; Decimal currency; Decimal fractions; Implemented curriculum; Intended curriculum; Isaac Greenwood; Metric system; Nicolas Pike; Thomas Dilworth; Vulgar fractions

   This chapter will examine the extent to which money calculations, and calculations concerned with weights and measures, as well as related problem-solving tasks, formed part of the elementary arithmetic curriculum in the North American European colonies during the period 1607–1775, and especially during the eighteenth century. The main analyses will be of intended and implemented arithmetic curricula in the colonies in the seventeenth and eighteenth centuries. It will be assumed that data relating to the author-intended curriculum can be found by examining arithmetic textbooks used in the schools, and data related to the teacher-implemented curriculum can be gathered by studying entries in North American cyphering books prepared by students in schools (Ellerton & Clements, 2012, 2014).

   Throughout the eighteenth century, school mathematics in the 13 British colonies was very different from what it is in North America in the twentieth-first century. On that issue, Ellerton and Clements (2012) provided data which supported the following strong statement:

> Two centuries ago, American teachers did not stand at the front of the room and teach, and most students, even those studying mathematics, did not own a mathematics textbook. Written examinations of any kind were not used. Most teachers of any branch of mathematics did not have formal qualifications in mathematics. (p. 2)

Ellerton and Clements (2012) summarized a cyphering tradition in which school arithmetic curricula for learners who were 10 years of age, or more than that, were based on an *abbaco* sequence that had probably originated in India and in Arab nations in the Middle Ages, and had been taken up by European merchants from about the twelfth century onward (Radford, 2003; Smith & Karpinski, 1911). That business-oriented curriculum was translated into

North America by European settlers in the seventeenth century. With three exceptions—specifically, textbooks by Isaac Greenwood (1729), Pieter Venema (1730), and Theophilus Grew (1758)—school mathematics textbooks used in the North American colonies were written by European (mainly British) authors, and money problems were couched in European currency terms, especially "sterling"—British pounds, shillings, pence and farthings. In the colonies this same British nomenclature was used, but the value of a pound in one colony often differed from the value of a pound in another colony. In a similar way, problems on weights and measures were based on European units—and in most cases, British units.

In fact, Greenwood's (1729) text did not achieve a wide circulation, and although occasionally an arithmetic textbook written by a German, or a Dutch, author, might be found in a school, most textbooks were by British authors. But, as indicated above, it should not be assumed that implemented arithmetic curricula in North American schools can be identified from analyses of textbooks, because most students—and probably most teachers—never owned, or rarely ever saw, an arithmetic textbook (Burton, 1833; Ellerton & Clements, 2012; Wickersham, 1886). That said, there can be little doubt that there was an overriding commercial thrust in *abbaco* curricula which had emerged from European reckoning schools in which tasks involving money, and weights and measures, were of central importance.

Patricia Cline Cohen (1982) questioned the relevance of the content and forms of mathematics that school students in the North American colonies were expected to learn. She wrote:

> Arithmetic was a commercial subject through and through and was therefore burdened with the denominations of commerce. Addition was not merely simple addition with abstract numbers, it was the art of summing up compound numbers in many denominations—pounds, shillings, pence; gallons, quarts, and pints (differing in volume depending on the substance being measured), acres and rods, pounds and ounces (both troy and avoirdupois), firkins and barrels, and so on. Eighteenth-century cyphering books show that students had to memorize all these tables of equivalences before embarking on the basic rules and operations. A large chunk of time was spent on a subject called "reduction"—learning to reduce a compound number to its smallest unit in order to facilitate calculations. Students would practice on questions like "How many seconds since the creation of the world?" and "How many inches in 3 furlongs and 58 yards?" This would prepare them for more advanced problems, such as "What will ten pairs of shoes cost at 25s 6d a pair?" (p. 121)

In this chapter data from textbooks and from cyphering books will be presented and analyzed, and that analysis should enable Cohen's claims to be evaluated.

What Mathematics Did North American Students Study in Schools Before 1792?

The *Abbaco* Heritage

A curious feature relating to school arithmetic in North America during the eighteenth century is that for the most part it was not the product of careful intentional design by academic mathematicians, or even by education administrators attempting to prescribe what students ought to learn in schools within their jurisdiction. Rather, it was the product of centuries of evolution, with its elements having originally been conceived by arithmetically-inclined practitioners seeking to improve profit margins in local, national and international trading ventures (Grendler, 1989). According to one tradition, it was Leonardo of Pisa (better known as "Fibonacci") who, around the year 1200 CE, wrote the treatise *Liber Abbaci* in which, among other things, he summarized the commercial arithmetic practices of Arab merchants that he had learned while living with his father in the present-day Algerian town of Béjaïa, formerly known as Bugia (Devlin, 2011; Gies & Gies, 1969; Høyrup, 2014).

Other scholars have attributed the original inflow of Hindu-Arabic arithmetic and practices into Europe to sources located in Southern Europe. Richard W. Hadden (1994), in *On the Shoulders of Merchants,* argued that *abbaco* arithmetic had its origins "outside the mainstream of mathematics" (p. 87), and that it is likely that modern mathematics "owes its origination to 'counting-house' needs of 'money-changers' of Lombardy and the Levant" (from Bochner (1966), quoted by Hadden (1994), p. 88).

The merchants' desire to predict and control profits resulted in the need to develop a class of so-called "algorists" who could analyze and solve the problems of mercantile arithmetic. Thus, for example, in 1613 Richard Witt systematized the calculation of compound interest (Lewin, 1970, 1981), and reckoning masters were expected to learn and apply Witt's system accurately. If a new approach appeared to be promising then it would be adopted by practitioners and authors of arithmetics. Thus, the process of curriculum innovation continued via a process of evolution. But, publication of a method in a book, no matter how prestigious the author or the publisher, was no guarantee that this new approach would be used by teachers in writing schools or apprenticeship schools.

According to Jens Høyrup (2008), *abbacus* schools, which operated in Italy from the thirteenth century, were

> … primarily frequented by merchant and artisan youth for … two years (around the age of 11), who were taught the mathematics needed for commercial life: calculation with the Hindu-Arabic numerals; the rule of three; how to deal with the complicated metrological and monetary systems; alloying; partnership; simple and composite discount; the use of "single false position"; and area computation. Smaller towns might employ a master; in towns like Florence and Venice private *abbacus* schools could flourish. In both situations *abbacus* masters had to compete, either for communal positions or for the enrolment of students. (pp. 4–5)

The period 1200–1800 was one of mercantile expansion and competition, with "reckoners," or "scriveners," being increasingly found in all major mercantile centers across Southern Europe (Radford, 2003; Van Egmond, 1980). Arithmetically-capable 10 or 11 year-old boys were likely to be bound as apprentice scriveners, and commercial pressures meant that a form of money-based, trade-based "school" arithmetic developed in writing and apprenticeship schools. During the period 1607–1776 this *abbaco* arithmetic was transplanted across the Atlantic and into the American colonies, and therefore school arithmetic in the colonies came to be heavily influenced by the needs and practices of merchants (Ellerton & Clements, 2012). Methods for solving classes of problems were carefully recorded in students' cyphering books, which served as pseudo-textbooks for generations of teachers.

There was no guarantee that a major mathematical breakthrough—such as François Viète's (1579) and Simon Stevin's (1585) expositions of decimal arithmetic, or Isaac Newton's and Gottfried Leibniz's development of calculus—would become part of school mathematics. For example, although Viète and Stevin conceptualized decimal fractions in the late sixteenth century and, in the case of Stevin, showed how these could be usefully applied in money tasks, and in problems involving weights and measures, during the seventeenth and eighteenth centuries their recommendations only occasionally found expression in the curricula of North American schools (Ellerton & Clements, 2014).

The most elementary forms of the *abbaco* curriculum began with a brief discussion of assumptions behind the numeration system in which the Hindu-Arabic digits (1, 2, 3, 4, 5, 6, 7, 8, 9 and 0) were used, and a place-value notation was adopted. In such a system the symbol "906" was used to denote the number "nine hundred and six" for which there were nine 100s, no 10s, and six units (that is to say, $906 = 9 \times 10^2 + 0 \times 10^1 + 6$). However, this "powers-of-ten" elaboration was not always to be found in commercial textbooks or in handwritten cyphering books. Then would follow a statement on what it meant to add two numbers, and an algorithm based on the Hindu-Arabic system would be stated as a "rule" or as a "case." After that, examples were shown which illustrated how and when that algorithm was to be used. Students would then be invited to apply the algorithm in written exercises. The same process would be repeated for subtraction, multiplication and division of whole numbers. Sometimes the algorithms would change—for example, in the seventeenth century the famous "scratch" or "galley" algorithm for division gave way to the simpler short- and long-division algorithms that are still used today (Ellerton & Clements, 2014).

The sections on numeration and on the four operations would be followed by sections on "compound operations" in which algorithms would be presented related to money and to measures of everyday quantities like length, area, volume, capacity, weight, time and angle. These applied algorithms would enable quantities to be added, subtracted, and sometimes multiplied and divided. Then would come sections on "reduction" and the several "rules of three." In both of those sections, the relationships which had been introduced in the "compound operations" section would be heavily utilized.

Of special interest for this present book is the extent to which the study of decimal fractions was introduced into the elementary *abbaco* curriculum sequences in North American schools during the period 1607 through 1810.

Author-Intended Curricula, as Interpreted Through Textbook Emphases

In order to ascertain key components of author-intended school arithmetic curricula within those North American British colonies that would, in 1775, become the United States of America, a content analysis was carried out with respect to compound operations, for five textbooks used in the schools in the eighteenth century. The books chosen for content analysis, listed in chronological order of publication, were:

- Isaac Greenwood (1729). *Arithmetick vulgar and decimal.* … Boston, MA: S. Kneeland and T. Green
- Thomas Dilworth (1762). *The schoolmasters assistant* … London, United Kingdom: Henry Kent
- John Hill (1772). *Arithmetick, both in the theory and practice* … London, United Kingdom: W. Strahan
- Nicolas Pike (1788). *A new and complete system of arithmetic, composed for the use of the citizens of the United States.* Newbury-Port, MA: John Mycall
- Benjamin Workman (1789). *The American accountant or schoolmaster's new assistant* … Philadelphia, PA: John M'Culloch

Three of these five texts—those by Greenwood, Pike, and Workman—were written in North America. Although Workman lived in Pennsylvania, and his book was published in Pennsylvania, it was largely based on an arithmetic written in Ireland, for Irish students, by John Gough. Gough typically referred to units of money and weights and measures, and emphasized problem scenarios which would have been recognized by Irish students but not necessarily North American students. The texts by Hill and Dilworth were written in England, presumably with British students in mind.

Although the five textbooks by Greenwood, Dilworth, Hill, Pike, and Workman were certainly not the only arithmetic texts found in some schools in the 13 colonies, each was used in the colonies. When analyzing the texts we focused on the following 12 topics or themes:

1. *Notation or numeration:* Discussion of this theme always occupied the first few pages of an arithmetic textbook. The purpose was to explain the basis of the Hindu-Arabic, place-value notation for natural numbers.

2. *Four operations tasks, free of context*: This always followed immediately after the notation or numeration section. Rules for adding, subtracting, multiplying and dividing natural numbers were elaborated, examples shown, and exercises set.

3. *Money calculations.* Methods for adding, subtracting, multiplying and dividing sums of money (usually expressed as sterling currency—pounds, shillings, pence and farthings) were given, examples shown, and exercises set.

4. *Weights and measures calculations:* Methods for adding, subtracting, multiplying and dividing weights and measures of quantities were given, examples shown, and exercises set.

5. *Reduction tasks involving money*: Methods were given for dealing with tasks in which a sum of money was to be expressed as the total number of smaller units in that sum (e.g., "How many farthings in 4 pounds, 13 shillings and 9 pence?"), or the reverse (e.g., "Suppose I have 47,211 pence—express that amount in pounds, shillings and pence"); then illustrative examples were shown, and exercises set.

6. *Reduction tasks involving weights and measures (not money)*: Methods were given for dealing with tasks in which a quantity was to be expressed as the total number of smaller units in that quantity (e.g., "How many seconds elapse between 12 o'clock midday and 12 o'clock midnight?"), or the reverse (e.g., "How many hours, minutes and seconds are there in 47,211 seconds?").

7. *Rules of three:* Rules and methods were elaborated for dealing with proportion tasks for which it was intended that one of the various rules of three—the direct rule of three, the indirect or inverse rule of three, the double rule of three, etc.—would be applied.

8. *Practice:* Most books included "clever" rules and methods based on algorithms developed over centuries that could expedite money calculations.

9. *Other money topics:* These "other" topics included tare and tret, currency exchange, loss and gain, discount, rebate barter, simple interest, compound interest, equation of payments, and annuities.

10. *More advanced topics:* Included here were mensuration, arithmetical progressions, geometrical progressions, involution and evolution (including square and cube roots), mensuration of plane and solid figures, permutations and combinations.

11. *Vulgar (or common) fractions:* Note that the material following a heading like "Direct Rule of Three with Fractions" was regarded as belonging to this category.

12. *Decimal fractions:* Note that the material following a heading like "Direct Rule of Three with Decimals" was regarded as belonging to this category.

Analysis of the 12 themes in the five textbooks has produced the percentages of pages dedicated to those themes shown in Table 2.1.

Table 2.1
Summary of Mathematical Topics in Five Arithmetic Textbooks Used in North American Schools, 1729–1789

Text-book Author	Numer-ation (%)	Four Operat-ions (%)	Money Calcu-lations (%)	W&M Calcu-lations (%)	Reduc-tion: Money (%)	Reduc-tion: W&M (%)	Rules of Three (%)	Prac-tice (%)	Other Money Topics (%)	Other Higher Topics (%)	Vulgar Frac-tions (%)	Dec. Frac-tions (%)
Green-wood (1729)	2	10	1	4	2	1	11	9	26	11	8	15
Dilworth (1762)	2	3	4	12	2	2	12	7	26	16	8	7
Hill (1772)	2	4	2	2	5	3	10	4	31	17	4	15
Pike (1788)	1	5	5	8	4	4	8	4	24	16	7	14
Workman (1789)	2	9	7	9	2	3	15	8	25	6	7	7
Overall %	2	6	4	7	3	3	11	6	26	13	7	12

The Teacher-Implemented Arithmetic Curriculum

Entries in Table 2.2 were arrived at as a result of an analysis of the topics covered, and the percentages of pages dedicated to those topics, in 21 North American arithmetic cyphering books prepared between 1742 and 1791. Each of the cyphering books is in the Ellerton-Clements collection. The upper limit of 1791 was chosen because that was the year before a national mint was established. Throughout the period 1740–1791 no distinctly North American coins (of the dollars and cents variety) were minted. Note that any cyphering book in the Ellerton-Clements collection which focused on navigation or surveying was not included in the set of books analyzed for the purposes of Table 2.2.

Vulgar (or Common) Fractions and Decimal Fractions in Intended and Implemented Arithmetic Curricula

A consideration of entries in the last two columns of Tables 2.1 and 2.2 prompted two important observations with respect to vulgar and decimal fractions in the intended and implemented curricula.

Table 2.2

Summary of Mathematical Topics in 21 North American Cyphering Books, 1742–1791

Name and Year(s)	Numer-ation (%)	Four Operat-ions (%)	Money Calcu-lations (%)	W&M Calcu-lations (%)	Reduc-tion: Money (%)	Reduc-tion: W & M (%)	Rules of Three (%)	Prac-tice (%)	Other Money Topics (%)	Other Higher Topics (%)	Vulgar Frac-tions (%)	Dec. Frac-tions (%)
Jonathan Livermore 1742–1743	0	24	0	29	5	24	18	0	0	0	0	0
Peter Tyson 1764–1767	1	11	9	24	11	9	21	6	7	0	0	0
Hulett Cornwell 1766	1	5	8	11	0	0	14	8	19	17	10	7
Sally Halsey c.1767	1	1	0	12	0	0	15	20	41	10	0	0
Paul Coolledge 1767–1772	1	9	2	14	2	3	23	8	21	15	1	1
David Townsend 1770–1774	0	0	0	1	0	2	8	8	64	17	0	0
John Grey 1771	0	7	1	2	7	7	36	7	18	16	0	0
Silas Mead 1772	1	10	7	21	9	9	18	16	5	3	0	0
Amos Lockwood 1774	2	41	10	46	0	0	0	0	0	0	0	0
Cornelius Houghtaling 1775	1	0	2	3	4	3	11	2	30	31	15	2
John Wright 1779	0	16	16	25	6	11	27	0	0	0	0	0
Loring Andrews c. 1780	1	12	2	11	3	1	10	7	14	7	21	10
Thomas Burlingame 1781–1782	1	10	7	17	11	9	31	9	4	0	0	0
Henry Wallentine 1783	0	40	7	17	21	12	2	0	0	0	0	0
John Anderson 1784–1785	1	13	16	20	4	12	10	7	13	3	0	0
John McDuffee 1786	0	19	21	10	4	12	33	2	0	0	0	0
Isaac Atkinson 1786–1787	0	0	0	0	0	0	64	23	9	5	0	0
Seth Torrey 1788	0	11	0	14	7	2	25	0	23	18	0	0

(continued)

Table 2.2 (continued)

Summary of Mathematical Topics in 21 North American Cyphering Books, 1742–1791

Name and Year(s)	Numer-ation (%)	Four Operat-ions (%)	Money Calcu-lations (%)	W&M Calcu-lations (%)	Reduc-tion: Money (%)	Reduc-tion: W & M (%)	Rules of Three (%)	Prac-tice (%)	Other Money Topics (%)	Other Higher Topics (%)	Vulgar Frac-tions (%)	Dec. Frac-tions (%)
Mahala Gove, 1788	0	19	3	36	6	22	0	0	0	0	0	14
Not Known, c.1790	0	2	0	5	9	2	19	18	45	0	0	0
Winthrop Dearborn 1791–1800	10	16	20	2	7	5	40	0	0	0	0	0
Mean %	1	12	6	15	6	7	21	7	15	6	2	2

The first observation was that all five textbooks examined (see Table 2.1) included educationally significant sections on both vulgar and decimal fractions; and second, of the 20 students who prepared the cyphering books on which Table 2.2 was based, only five prepared pages, in their cyphering books, on either vulgar fractions or decimal fractions. Hulett Cornwell (1766), Paul Coolledge (1767–1772) and Loring Andrews (c. 1789) included sections on both types of fractions; Cornelius Houghtaling (1775) included a large section on vulgar fractions and a small section on decimal fractions; and Mahala Gove (1788), included a relatively large section on decimal fractions but nothing on vulgar fractions.

Toward the end of the sixteenth century both François Viète (1579) and Simon Stevin (1585) stressed how the use of decimal fractions could simplify the carrying out of lengthy multiplications or divisions. Since traditional *abbaco* arithmetic dealt with multiplication and division of whole numbers at an early stage, one might have expected the lag time, before decimal fractions were introduced into schools, to be short. In fact, reasonably early in the seventeenth century some authors of arithmetic textbooks for schools incorporated both decimal fractions and logarithms into their intended arithmetic curricula.

Following publications by John Napier (1614, 1619) and Henry Briggs (1617), Edmund Wingate (1624), an English mathematician temporarily based in Paris, emphasized the power of the combination of decimal fractions and common logarithms—that is to say, logarithms to the base 10—to assist practitioners, such as surveyors, navigators, and carpenters, to make the kinds of calculations that they were likely to need to make in their daily workplaces. On returning to England, Wingate (1630) boldly used the "decimal point" in his book *Of Natural and Artificiall Arithmetique* (Glaisher, 1873). which was aimed at schools. He distinguished between "natural or common arithmetick" and "artificial arithmetick"—the latter referring to the arithmetic of decimals and logarithms.

In the 1680s John Kersey (1683) claimed that before Wingate died he asked Kersey to revise *Arithmetique Made Easie* so that the new version would revert to the traditional *abbaco* arithmetic sequence—with numeration, the four operations, compound operations on money and weights and measures, reduction, practice, the rules of three, alligation, fellowship and false position—being initially developed through whole numbers. Then, later, the same topics should be developed through common—that is to say, vulgar—fractions, and later still, through decimal fractions (Ellerton & Clements, 2012). In Kersey's (1683) revised edition of Wingate's text, decimal fractions were not formally dealt with until Chapter 22. Kersey (1683) claimed that "decimal fractions are being commonly abused, by being applied to all manner of questions about money, weight, &c, when indeed many questions may be resolved with much more facility by vulgar arithmetic" (p. 168).

So, although Wingate clearly *intended* decimal fractions to become an integral part of school arithmetic, Kersey effectively gave permission to British teachers to continue to follow a non-decimal, traditional sequence of *abbaco* arithmetic. Other authors of arithmetics adopted a similar approach to Kersey's. Edward Cocker's (1677) arithmetic, published by John Hawkins, basically avoided decimal fractions, but in 1685 Hawkins published *Cocker's Decimal Arithmetick* which was intended to serve the needs of those who went beyond the first book. In his preface, Hawkins acknowledged that, in taking this approach of presenting, in a separate text, elementary *abbaco* arithmetic which do not use decimal fractions, he was following the lead of Kersey (Cocker, 1685, p. vi).

In fact, though, the 1685 first edition of *Cocker's Decimal Arithmetick* set out, with great clarity, what the future might hold so far as the applications of decimals were concerned. After showing how simple multiplication of decimals could be used to find the "content" of a table whose length was 18.75 feet and breadth 3.5 feet, Cocker (1685) wrote:

> Here by the way, take notice, that although amongst artificers the two foot rule is generally divided, each foot into 12 inches, &c., yet for him that at any time is employ'd in the practice of measuring, it would be most necessary for him to have his two foot rule, each foot divided into 10 equal parts, and each of those parts divided again into 10 other equal parts: so would the whole foot be divided into 100 equal parts, and thereby would it be made fit to take the dimensions of any thing whatsoever, in feet and decimal parts of a foot; and thereby the content of any thing may be found exactly, if not more exactly and near, than if the foot were divided into inches, quarters, and half quarters. (p. 45)

Cocker proceeded to demonstrate how much more cumbersome the calculation of the "content" of a table 18 feet 9 inches long and 3 feet 6 inches wide would be if the "normal" method were used—18 feet 9 inches would be converted to 225 inches and the 3 feet 6 inches to 42 inches. Then, after multiplying 225 by 42, the product 9450 would be obtained, which after division by 144 would give $65\frac{90}{144}$; then, that mixed fraction would need to be interpreted in relation to the original problem. Cocker (1685) described that method as "tedious" when compared with "the decimal way" (p. 47).

It was one thing, however, to present good reasons for the introduction of decimal arithmetic into schools and another thing for teachers to adopt that innovation. The longstanding *abbaco* tradition influenced how parents, teachers and merchants thought about school arithmetic, and teachers needed more than mere argument to be persuaded to depart from such time-honored, familiar practices. Cocker's (1677) traditional arithmetic would prove to be far more popular than his *Decimal Arithmetick,* with revised editions of the non-decimal text being published for the next 150 years. *Cocker's Decimal* Arithmetic would also be reprinted, but it was always much less popular than the traditional arithmetic, and that was especially true in the North American colonies. Cocker's non-decimal arithmetic was imported into North America, and much used there, but there is no evidence that his *Decimal Arithmetick* was often used outside of Great Britain (Karpinski, 1980).

Those entries pertaining to vulgar and decimal arithmetic in the right-hand columns of Tables 2.1 and 2.2 make it clear that historians of curriculum should never assume that the mere existence of chapters on a topic in popular textbooks implies that most students studied that topic. Some students might not have used a textbook at all, and others might have used a textbook but avoided the chapter(s) dealing with the topic. In other words, the implemented curriculum may not necessarily be the same as an author-intended curriculum.

Which Was Easier, the Author-Intended or the Implemented Arithmetic Curricula?

After carrying out the analyses which generated Tables 2.1 and 2.2, we then assumed that half of the "rules of three" tasks were concerned with money, and half with weights and measures. Working with that assumption, we calculated estimates of the percentage of pages in the textbooks and in the cyphering books that were dedicated to the arithmetic of money— by adding the mean percentage entries for money calculations, reduction tasks involving money, (rules of three tasks)/2, practice, and other topics predominantly requiring money calculations. On doing this, we found that the percentages of pages directly concerned with elementary rules, cases, calculations, model examples, and exercises associated with money were 44.5%, for textbooks, and also 44.5% for cyphering books. We concluded that in both the textbooks and the cyphering books almost one-half of the pages were dedicated to forms of arithmetic associated with money. In almost every case, the money calculations involved amounts of sterling currency expressed in pounds, shillings, pence and farthings—the only exception being with the topic of "Exchange" (which dealt with currencies in different nations).

By adding the mean percentages for W&M (weights and measures) calculations, reduction tasks involving weights and measures, and (rules of three tasks)/2, we obtained 15.5% for textbooks and 32.5% for cyphering books. We concluded that teacher-implemented curricula in North American schools during the period 1742–1792 were more concerned with weights and measures than were author-intended curricula.

Taken together, entries on money or weights and measures accounted for about 60% of pages in the textbooks, and 77% of pages in the cyphering books. The textbooks contained a higher percentage of pages concerned with higher-order arithmetic than the cyphering books (13% as opposed to 7% respectively). Correspondingly, a smaller percentage of the textbook pages (25%) dealt with easier topics (notation and numeration, four operations on natural numbers, money or weights and measures calculations, and reduction) than for the cyphering book pages (45%). Generally speaking, implemented curricula, as evidenced in the cyphering books, were easier than the author-intended curricula of the textbooks. Given the lack of arithmetical expertise of many teachers, that was not an unexpected finding (Ellerton & Clements, 2012). Only about 8% of pages in textbooks and 12% of pages in cyphering books were dedicated to numeration or to the four operations (without reference to money or weights and measures). There was a decidedly practical emphasis in the cyphering books.

The *Abbaco* Arithmetic Curriculum in North America in the Eighteenth Century

In order to give readers a more complete description of the *abbaco* curriculum which more or less controlled the study of arithmetic in colonial schools in the eighteenth century 12 pages from cyphering books will now be reproduced. Each page will appear on the left side of this book, and the right side will provide commentary on that page. Each page will correspond to a column heading in Table 2.2.

A Page Showing an Entry on Notation and Numeration (from Paul Coolledge's Cyphering Book, 1767)

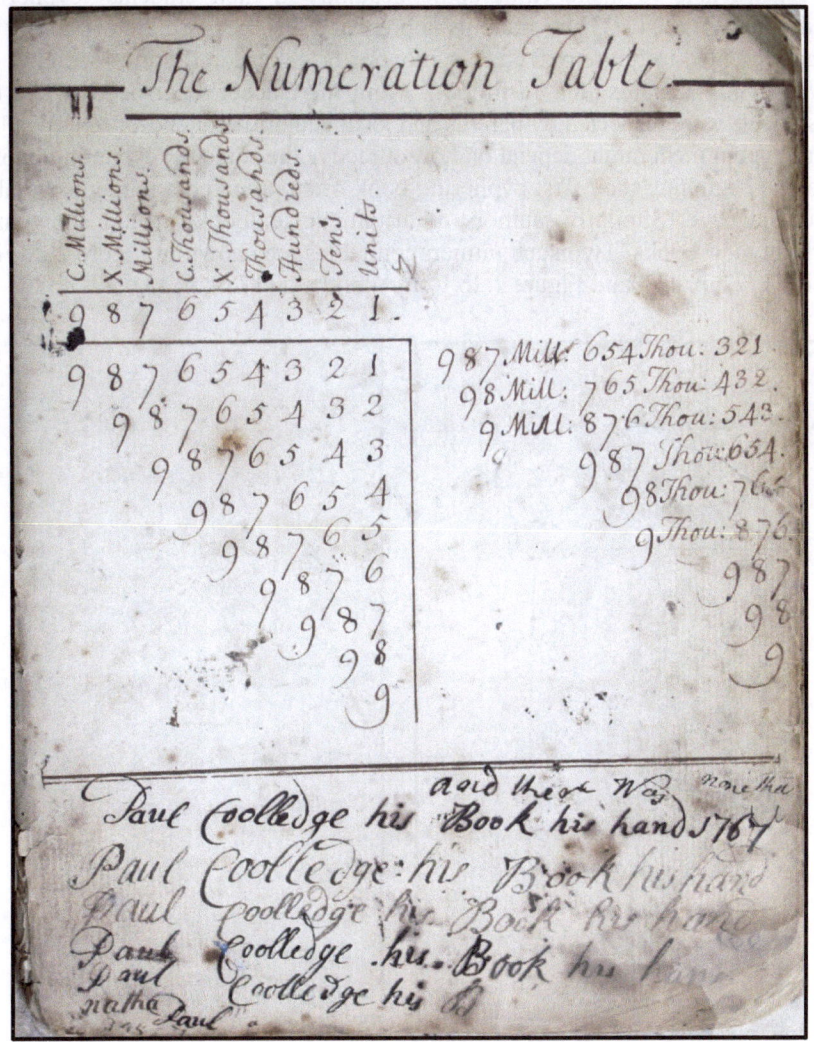

Figure 2.1. A numeration table in Paul Coolledge's (1767) cyphering book.

The page shown in Figure 2.1 was the first page in a cyphering book prepared by Paul Coolledge in the town of Weston, Massachusetts, in 1767. According to ancestry.com records, in the second half of the 1770s Paul would become a Revolutionary War soldier.

There is nothing particularly special about the page reproduced in Figure 2.1: indeed, first pages of cyphering books were often more attractive than this one. The numeration table, which went as far as hundreds of millions, was attractively presented. At the bottom of the page Paul wrote his name five or six times.

It should be obvious that the quality of penmanship at the bottom of the page did not match the quality of penmanship in the heading and in the actual numeration table. Almost certainly, that would have been because Paul's teacher wrote the upper two-thirds of the page (i.e., the part containing the numeration table) and Paul wrote the bottom one-third. It was common for teachers to make calligraphic headings in their students' cyphering books (Ellerton & Clements, 2014). The students appreciated this, because the cyphering book was likely to be presented to future potential employers, or college authorities deciding whether a student was fit for entry to their institution. From the teacher's perspective, the cyphering books of students were inspected by parents and local authorities at end-of-term displays, and a teacher's re-appointment might depend on how attractive the students' cyphering books were.

Students beginning their first cyphering book often prepared a numeration table on the first or second page. Similarly, authors of arithmetic textbooks would show numeration tables early in their books. Two such numeration tables are shown in Figure 2.2. Figure 2.2a is from Hill (1772, p. 13), and Figure 2.2b from Workman (1789, p. 13).

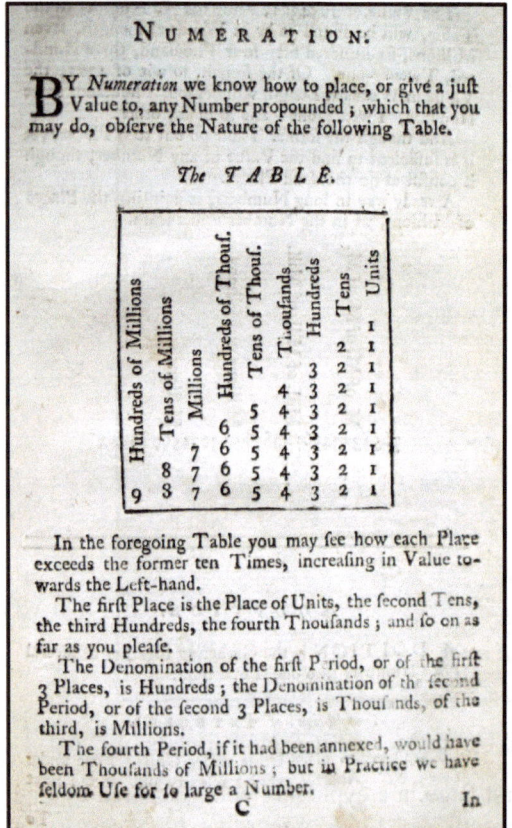
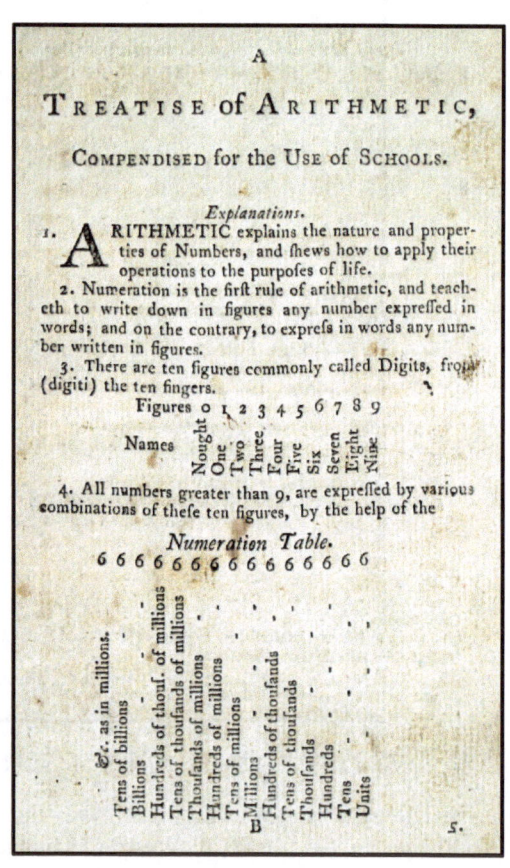

    (a) From Hill (1772, p. 13)   (b) From Workman (1789, p. 13)

Figure 2.2. Numeration tables—John Hill (1772) and Benjamin Workman (1789).

Loring Andrews' Book

Figure 2.1, which showed an early page from Paul Coolledge's (1767) cyphering book, and Figure 2.2, which shows pages from two arithmetic textbooks, illustrate the contrast between the different kind of arithmetical activity represented by cyphering books and commercially published textbooks. The actual arithmetic dealt with on the first page of Paul's book and on the two pages shown from John Hill's (1772) and Benjamin Workman's (1789) textbooks is quite similar—but Paul wanted to remind himself that this was his *personal* cyphering book. Beginning a cyphering book represented *rite de passage* for many 10-year-olds. By contrast, the textbook authors presented the content in a matter-of-fact way. Paul's teacher might have helped with the heading on the page, but at the bottom of the page Paul made it clear that beginning a cyphering book was a big event for him.

Another cyphering book summarized in Table 2.2 was that prepared by Loring Andrews, from Boston. Loring's manuscript is undated but the cover was crudely made out of a 1780-edition of the *Independent Chronicle and the Universal Advertiser*, a newspaper published in Boston, Massachusetts. Like Paul Coolledge, Loring was determined to celebrate his graduation to the world of cyphering. After a beautifully presented "Numeration Table," which went from "units" to "hundreds of millions," he wrote, "Loring Andrews his book" three times at the bottom of the first page. Between the numeration table and his proud declaration of ownership were two "aditions (*sic.*) of whole numbers"—Both of the additions involved four 7-digit whole numbers, which were added vertically. Checks were shown.

Loring's manuscript is now in very fragile condition, with the outside newspaper almost separated at the fold. But it is 235 years old, and its continued existence attests to the value placed on cyphering books, by Paul himself, by members of his family, and by his descendants. It is a very authentic eighteenth-century cyphering book, with 40 pages of rag paper (dimensions 11.75″ by 7.75″) stitched between the newspaper covers. The ink was of a typical dark brown variety. The standard of calligraphy and penmanship was only fair—but, almost certainly, it represented the best that young Loring could do. Topics covered were numeration, compound operations, addition, "substraction," multiplication, division, reduction, vulgar fractions, decimal fractions, rules of three, practice, simple interest, and fellowship. IRCEE and PCA genres were evident throughout.

From an educational point of view, however, there is something special about Loring's manuscript. It includes substantial sections on vulgar and decimal fractions—at a time when those topics were found in a minority of cyphering books. Perhaps, those topics were more likely to be studied in schools in larger towns and cities—like Boston, New York, or Philadelphia—than elsewhere.

A check of the entries in Table 2.2 will reveal that for Loring's manuscript there were higher *proportions* of pages on vulgar and decimal fractions than for any of the other 19 cyphering books covered. In the notes he wrote when introducing decimal fractions, Loring showed a second numeration table—this one not only included entries for numbers from "units" to "millions," but also numbers from "parts of 10" to "parts of 100 million." Paul's book included an emphasis on money and on measurement—which was consistent with the *abbaco* tradition. But, toward the end of his book, when solving problems, Paul used the vulgar and decimal fraction concepts that he had introduced earlier in his book.

Although around 1780, *some* students did study vulgar and decimal fractions, entries in Table 2.2 show that, despite the fact that almost all textbooks included sections on fractions, only a small proportion of students actually spent time trying to learn how to use them.

A Page Showing an Entry on Multiplication
(from Amos Lockwood's Cyphering Book, 1774)

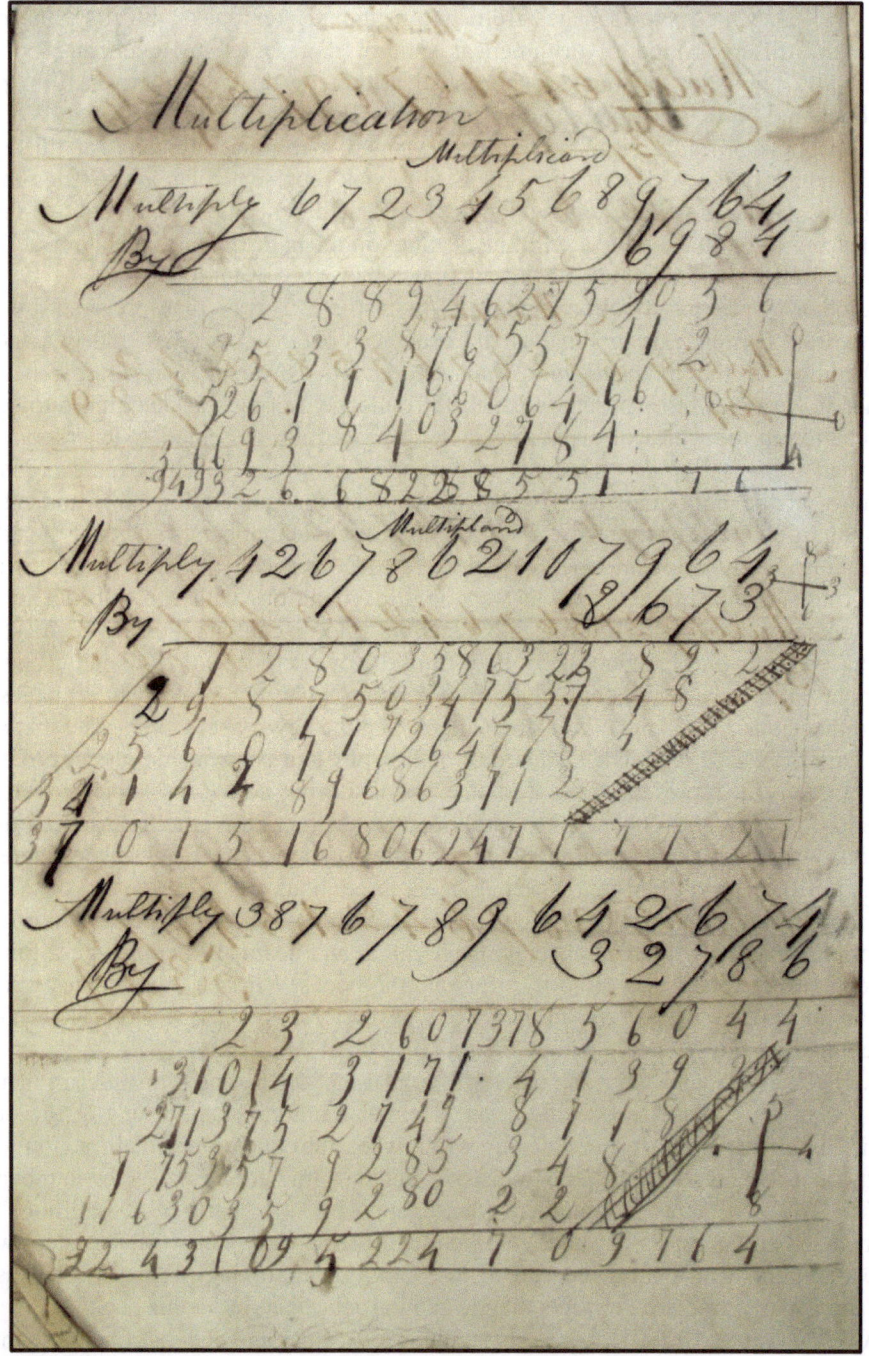

Figure 2.3. Multiplication, in Amos Lockwood's (1774) cyphering book.

Figure 2.3 shows the first page on multiplication in a cyphering book prepared in 1774 by Amos Lockwood, in the town of Warwick, Rhode Island. Although the penmanship is not particularly attractive, there are five noteworthy features:

1. Although Amos had just begun the topic of multiplication there were numerous pages in his cyphering book for which the multiplicand had at least 12 digits. In most examples the multiplier had at least three digits.
2. Note the use of the word "multiplicand" (on two occasions) to indicate the number that was being multiplied. It seems that in the seventeenth and eighteenth centuries the practice of naming the numbers involved in operations was common.
3. The answers given for most of the tasks were, in fact, wrong, yet …
4. Amos checked his answers using a "casting-out-nines" method (this procedure, which was used by Abraham Lincoln, is discussed in Ellerton & Clements, 2014). Thus, for example, the sum of the digits in 672345689764 is 67, which when divided by 9 has 4 remainder; the sum of the digits in 6984 is 27, which when divided by 9 has 0 remainder; the product of the remainders is 0, which when divided by 9 has 0 remainder; the sum of the digits in the answer which Amos gave was 77, which when divided by 9 had a remainder of 5. Thus, Amos checked his incorrect answer, and he concluded it was correct even though it was not.
5. A clue to why incorrect answers were obtained can be found by looking carefully at the quality and maturity of penmanship evident in the tasks. Almost certainly, an adult set the tasks, but numerals for the calculations were entered by Amos. The teacher did not check Amos's calculations carefully—he was probably convinced that answers were correct because casting-out-nines checks were shown.

This cyphering book was created immediately before and during the period of the Revolutionary War with Great Britain, and on its last page there is an important, handwritten tax protest letter signed by William Ellery (a signer of the Declaration of Independence), Joseph Wanton, Henry Ward, John Collins and John Mawdsley all of Newport, Rhode Island. This protest letter is headed: "Newport January 15th 1774." It began:

Gentlemen, in pursuance of an order of the Town, we inclose (sic.) you the Resolutions entered into, *Nemine contradicente* in a very full meeting here on Wednesday last. The attempts of the Ministry to establish an unlimited power in the British Parliament of levying taxes upon his Majesty's subjects in America at pleasure and the necessity of a firm union among the Colonies to prevent a Measure so utterly subversive of all our just rights are so obvious, that we shall not enlarge on the subject. We request you to lay these resolutions before a meeting of your Town as soon as possible and hope that such a union may take place as will enable us by the blessings of God to preserve our just rights and Privileges.

This letter was addressed to "The Worshipful Town Council of Warwick."

It is likely that the original document, signed individually by William Ellery, Joseph Wanton, Henry Ward, John Collins and John Mawdsley, was copied by hand numerous times (in unlikely places, such as the rear page of a cyphering book) and then forwarded surreptitiously to parents and other residents in various townships. In 1774, Rhode Island was still under British rule, and anyone found in the possession of such a document could have been in serious trouble. Another copy of the document (with the same five signatures) is now held in the Johnston Town Records Collection for 1759–1889 (Catalog number: MSS 202), and is recorded by the Manuscripts Division of the Rhode Island Historical Society.

Multiplying Money: From Peter Tyson's Cyphering Book (1764)

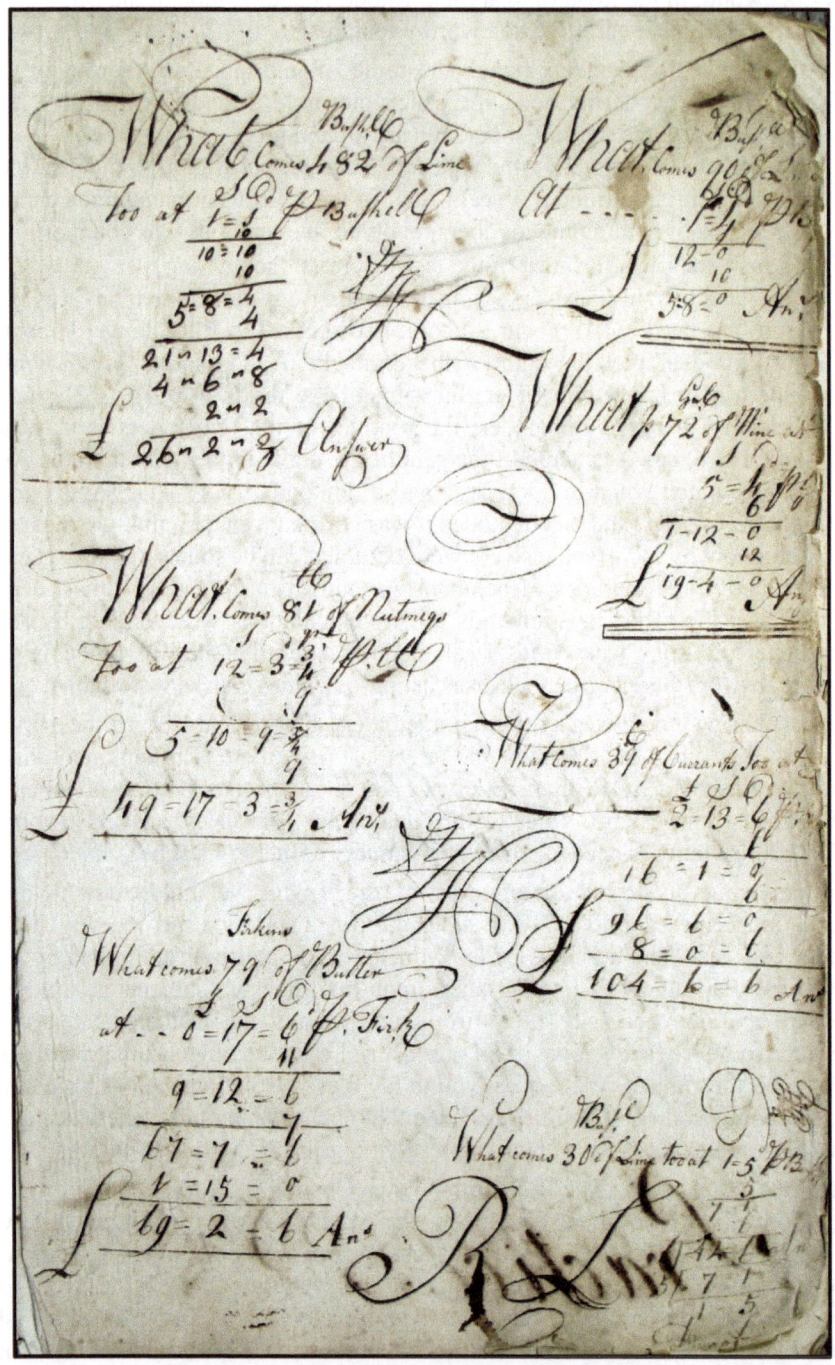

Figure 2.4. Multiplication of money, in Peter Tyson's (1764) cyphering book.

Analysis of Peter Tyson's Approach to Multiplying Money (Expressed in Sterling)

There are seven "multiplication-of-money" tasks shown in Figure 2.4. It is not clear what the calligraphic initials "$\mathcal{R}$ $\mathcal{L}$" represent at the bottom right corner, but there can be no doubt that this was Peter Tyson's cyphering book for his name appeared on numerous pages throughout the book. The calculations shown on the page were probably done by Peter Tyson without the aid of his teacher, but it is possible that "$\mathcal{R}$ $\mathcal{L}$" were the initials of a teacher.

It will be instructive to examine the two exercises shown at the top left of Figure 2.4. The wording for these tasks were:

Task 1: What comes to 482 bushels of lime at 1 shilling and 1 penny per bushel?
Task 2: What comes to 81 lb of nutmegs at 12 shillings, and 3¾ pence per lb?

Working for Task 1

Multiplication by 482 was achieved by multiplying by 400 and then adding the cost of 80 bushels and 2 bushels. Multiplication by 400 was achieved by multiplying by 10, then by 10 again, and then by 4. The method for calculating the cost of 80 bushels was not shown. Peter's use of a PCA ("Problem-Calculation-Answer") genre was evident.

Peter's setting out when multiplying 5 pounds, 8 shillings and 4 pence by 4 suggested that he used an algorithm for which the base was changed after each calculation, with the first calculation being 4 by 4 pence. Since, using base 12, 16 pence is 1 shilling and 4 pence, the 4 was "put down" and the 1 "carried." Then the 8 was multiplied by 4, to get 32, and the 1 added to get 33; then, since 33 shillings was 1 pound and 13 shillings (using a base 20), the 13 was ":put down" and the 1 "carried"; then 5 was multiplied by 4 to get 20, and the 1 added to get 21. And so, the product of 5 pounds, 8 shillings and 4 pence and 4 was calculated to be 21 pounds 13 shillings and 4 pence.

After a process of adding 21 pounds 13 shillings and 4 pence, 4 pounds 6 shillings and 8 pence, and 2 shillings and 2 pence, the final answer 26 pounds 2 shillings and 2 pence was obtained—but note that different bases also needed to be used in this addition.

Working for Task 2

Multiplication by 81 was achieved by multiplying by 9 and then by 9 again. As with the working shown for Task 1, different bases were used in each of these multiplications.

Having carefully examined all the cyphering books in the Ellerton-Clements collection, as well as many other North American cyphering books, we believe that, before 1792, whenever a North American student entered the working for a multiplication-of-money task into his or her cyphering book a "change-of-base" method similar to that adopted by Peter Tyson was used. That observation is important, given that in the 1780s Thomas Jefferson asserted that students used a different algorithm (see Chapter 3 in this book).

Addition of Weights and Measures: From John Grey's Cyphering Book (1771)

Figure 2.5. Some "weights and measures" calculations by John Grey (1771).

Analysis of John Grey's Approach to Adding Quantities When They Were Given in Written Form

The first page of John Grey's (1771) cyphering book has been reproduced in Figure 2.5. An examination of the binding and sewing for the cyphering book as it now exists suggests that it is likely that previous pages—probably showing a numeration table, the four operations on whole numbers, and operations on money—have been removed from the cyphering book.

John Grey prepared his cyphering book in Rhode Island which, in the 1770s was a bustling center of local and international trade. The first three of the pages of the cyphering book, as it now exists, show tasks in which additions were performed on cloth measure, liquid measure, dry measure, time, long measure, and land measure. Other cyphering books of the period show addition of other kinds of measures—including weight, volume (often referred to as "solid measure"), and angles (often called "circular measure"). Usually distinctions were made between avoirdupois, troy, and apothecaries measures for weight, and John make those distinctions in the next section of his cyphering book (which was concerned with subtraction of quantities).

John's entries for four tasks have been reproduced in Figure 2.5. With the two tasks concerned with "Addition of Cloth Measure" there is a distinction between English Ells and Flemish Ells. After the first heading John wrote: "4 Nails make 1 Quarter of a Yard; 4 Quarters 1 Yard; 3 Quarters make 1 Ell Flemish; 5 Quarters make 1 Ell English."

Thus, with the first task, starting at the right column, 18 Nails are obtained, so put down 2 Nails and "carry" 4 Quarters. Then, including the 4 being carried, one gets 32 Quarters, which is 6 English Ells and 2 left over. Then, with the left column, including the 6 carried, one should have got 7604, but in fact John recorded 7614. John then showed a check, which seemed to confirm his calculations had been correct (which, in fact, was not the case).

John's working for the Flemish Ells task was correct. With his check he added all the given quantities except the first (to obtain 5915 Flemish Ells, 2 Quarters and 1 Nail), and then this sum was added to the first quantity 391 Flemish Ells, 1 Quarters and 1 Nail), which gave the correct 6307 Flemish Ells, 0 Quarters and 2 Nails.

John used a structurally similar approach when tackling the two "liquid measure" tasks shown at the bottom of Figure 2.5. After stating the relevant relationships ("2 Pints make 1 Quart; 4 Quarts make 1 Gallon; 63 Gallons make 1 Hogshead; and 4 Hogsheads make 1 Ton"), John then showed two lengthy additions of the quantities. All of the calculations and checks were correct.

Analysis of the cyphering book data does not make clear whether John was expected to remember all of the relationships between quantities. The addition of a large number of quantities (like those shown in Figure 2.5) might have been useful for anyone working in a trading center like Providence (Rhode Island) or Newport (Rhode Island), and therefore could have been of relevance to John.

Part of the cyphering-book tradition was that all entries in cyphering books should be correct—after all, the book was intended to be a reference book for life. However, in an age when there were no automatic aids to calculation, it was understandable that, even when the teacher was conscientious, minor errors such as those which occurred with the task shown at the top left of Figure 2.5 could have found their way into cyphering books.

Expressing Money in Different Units: From John Anderson's Cyphering Book (1785)

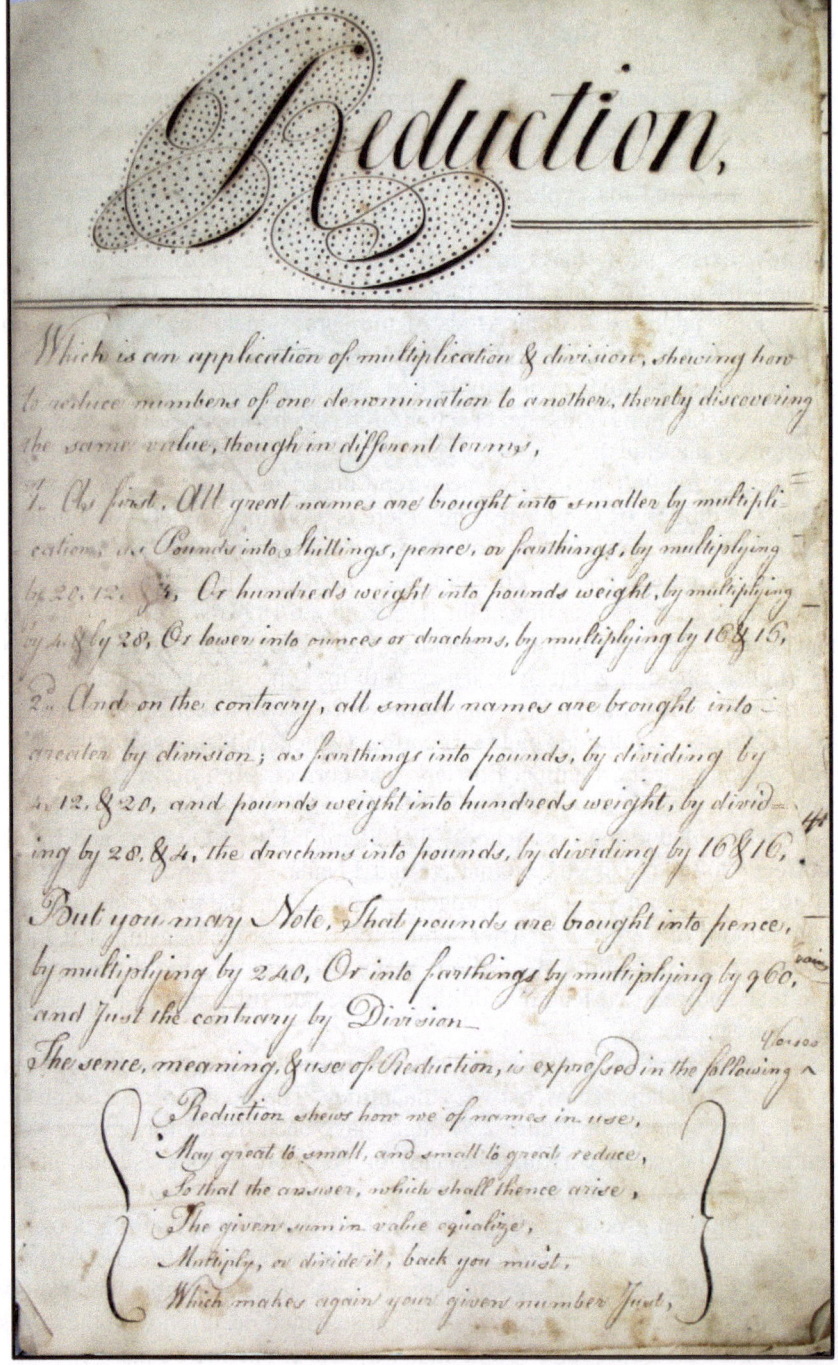

Figure 2.6. Notes on reduction of money, by John Anderson (1785).

Comments on John Anderson's (1785) Notes on Reduction of Money

Figure 2.6 is a reproduction of the notes on reduction that John Anderson made when preparing his cyphering book in Virginia in 1785. The quality of the calligraphic heading and the penmanship was high, and the verse at the bottom of the page was typical of verses that often appeared in cyphering books. Many students regarded their cyphering books as very personal documents, and standard entries on arithmetic were sometimes embellished with little poems and illustrations (Ellerton & Clements, 2012; 2014; Denniss, 2012; Wardhaugh, 2012).

Here we shall discuss an example of reduction that John included on the page in his cyphering book. This example was on the page following the one reproduced in Figure 2.6.

The task was:

> In 325 pounds 15 shillings and $7\frac{1}{2}$ pence, how many farthings?

This is how John set out his solution (on the left), together with a "proof" (on the right):

In 312750 farthings, how many pounds, shillings, pence and farthings?

£	s	d
325	15	$7\frac{1}{2}$

$$\underline{20}$$

6515

$$\underline{12}$$

78187

$$\underline{4}$$

312750

$4\langle\underline{312750}$

$12\langle\underline{78187} - \frac{1}{2}$

$2/0\langle\underline{651/5} - 7$

$£325..15..7\frac{1}{2}$ Proof

Although the notes shown in Figure 2.6 suggest that mere multiplication and division of money was involved in this working, that was not the case. Consider, for example, the conversion of pounds, shillings and pence (on the left). One started multiplying by 20, and that multiplication applied to the 325, which was on the left. By contrast, with the algorithmic procedure for multiplication of money sums expressed in pounds, shillings and pence, one began on the right side of the given amount; second, the 15 (shillings) were added in *before* the next multiplication (by 12)—that contrasted with the algorithm for the multiplication of money; third, the 7 (pence) were added in *before* the next multiplication (by 4); and fourth, in the final line, 2 farthings (i.e., $\frac{1}{2}$ pence) were added after the previous multiplication by 4. The reverse procedure (which constituted the "proof") differed from the standard algorithm for dividing pounds, shillings, pence and farthings.

A Page Showing Double Rule-of-Three Calculations by Silas Mead (1772)

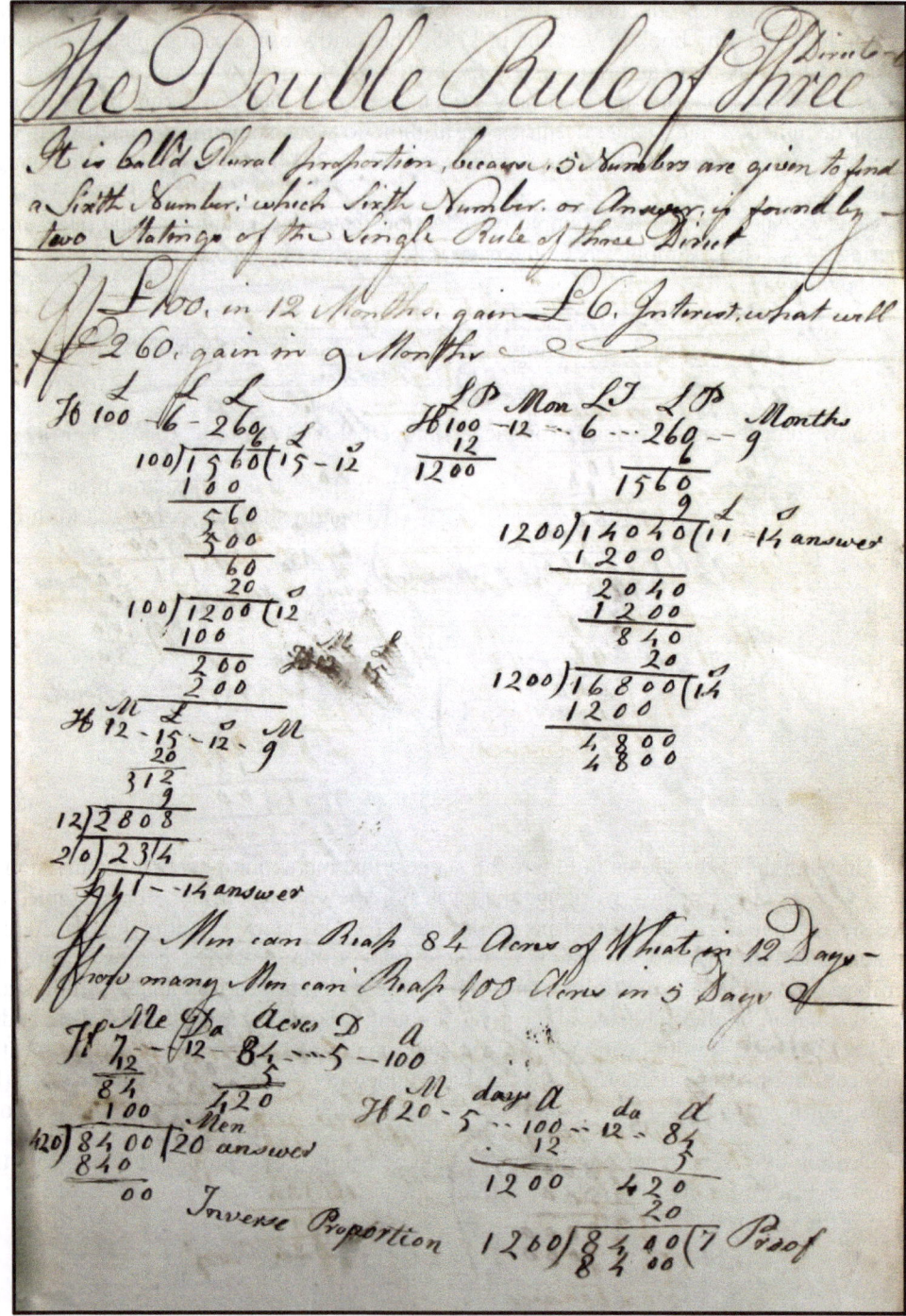

Figure 2.7. Double rule-of-three calculations by Silas Mead (1772).

Comments on Silas Mead's (1772) Calculations for a Double Rule-of-Three Task

Figure 2.7 is a reproduction of the notes on the double rule-of-three which Silas Mead made when preparing his cyphering book in Connecticut in 1772. Other pages in the same cyphering book dealt with the "direct rule of three," the "rule of three inverse," and the "double rule of three inverse." According to the notes in Silas's book:

> *The rule of three (direct)* is the rule of proportion because it shows what relation or proportion one number bears to another—like when three numbers are given to find a fourth, which is the answer. Also, multiply the second number by the third and divide by the first.

> *The rule of three (inverse)* is contrary to the rule of three direct. For, as in the rule of three direct, you multiply your second number by your third, and divide by your first, here you multiply your first by your second, and divide by your third, and the fourth is the answer.

> *The double rule of three direct* is called plural proportion because 5 numbers are given to find a sixth; which sixth number or answer is found by two statings of the single rule of three direct.

> *The double rule of three inverse.* The rule … is as follows: The three numbers belonging to the supposition must be your first, second and third numbers, and the demand is your fourth and fifth numbers.

The rules were vague. Each statement was followed by at least one model example (the two examples in Figure 2.7 were the only examples shown for the double rule of three direct).

An Example Showing the Double Rule of Three in Silas Mead's (1772) Cyphering Book

The first example was

If £100 in 12 months gain £6 interest what will £260 gain in 9 months?

This example shows how the concept of simple interest could be related to the double rule of three. The first thing Silas did was to find out what the interest on £260 would be for 12 months. By the first rule of three direct this was equal to £260 multiplied by 6 and then divided by 100. That working is shown in Figure 2.7, and the answer obtained was 15 pounds 12 shillings. Then a second application of the rule of three direct was used to find how much interest would be obtained in 9 months if 15 pounds 12 shillings were obtained in 12 months. This was achieved by multiplying 312 (shillings) by 9, to get 2808 (shillings) and then dividing this by 12, to get 234 (shillings) which, after dividing by 20, gave an answer of 11 pounds 14 shillings.

The rules of three were regarded as crucially important, and extremely difficult (Ellerton & Clements, 2012, 2014). The setting out in Silas Mead's (1772) cyphering book should make it clear why that would have been the case. A much more straightforward approach would have been to multiply £6 by the fraction $2\frac{3}{5}$, and then to multiply the result of this calculation by $\frac{3}{4}$. But cyphering-book evidence suggests that Silas had never studied vulgar fractions—and, although one cannot say for certain, the same might also have been true of his arithmetic teacher.

An Unknown Author (c. 1790) Uses the Rules of "Practice"

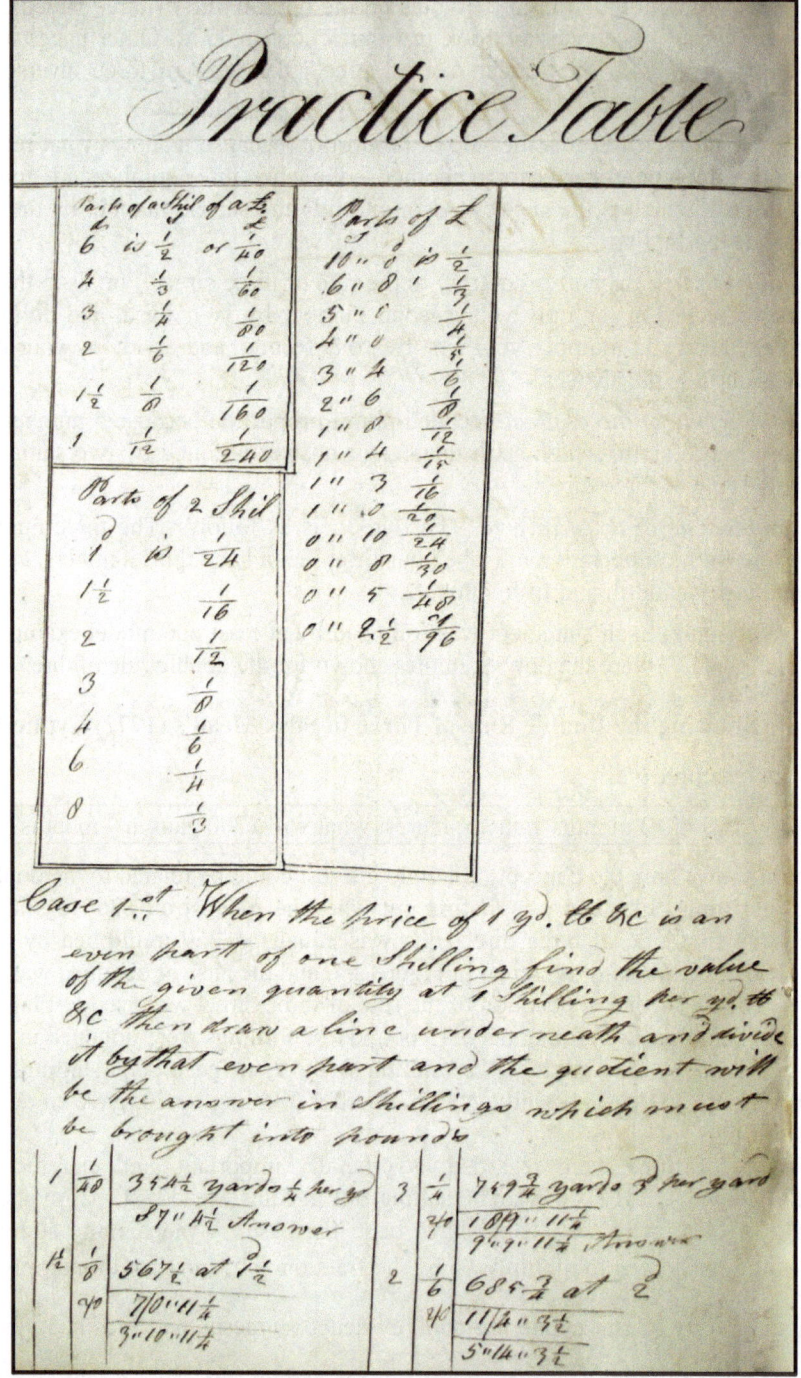

Figure 2.8. Rules of "practice" calculations by an unknown author (c. 1790).

"Practice" was an elementary part of the *abbaco* sequence and often appeared either immediately before or immediately after the rules of three. The method of practice relied upon knowing what would, in the twenty-first century, be regarded as unit fractions associated with quantities. Thus for example, the student was expected to learn that 6*d*, 4*d*, 3*d*, 2*d*, $1\frac{1}{2}d$, and 1*d* were $\frac{1}{2}, \frac{1}{3}, \frac{1}{4}, \frac{1}{6}, \frac{1}{8}$ *and* $\frac{1}{12}$ of a shilling, respectively. Similar "aliquot parts" of quantities (i.e., unit fractions for quantities) needed to be known for "parts of a tun" (e.g., 10 cwt was half a tun) and "parts of a cwt" (e.g., 14 lbs was half a quarter, and 4 quarters made a cwt).

Ellerton and Clements (2014) referred to the rules of practice as "magic" procedures (p. 81) which had been worked out by algorists to expedite elementary money calculations. Students were not expected to know the reasons behind the rules. In Figure 2.8, the following rule is shown:

> When the price is pence and [there is] no even part of a shilling find the value of the given quantity at 1 shilling per yard, divide the price into aliquot parts for division and the sum of the quotients arising from these will be the answer.

Exactly what this statement meant would not have been clear to most students—but seven examples were given in the hope of making the meaning clear. The first problem asked: "What will $487\frac{1}{2}$ yards cost at 3*d* per yard?" Applying the rule that $487\frac{1}{2}$ needed to be divided by 3 and by 4, the sum of the results of those two divisions was obtained, and then this sum (which was in shillings) was converted to pounds by dividing by 20, giving 10 pounds 3 shillings and $1\frac{1}{2}$ pence as the answer. PCA genre was adopted. The method which was used clearly fitted the description "rules without reason" (Skemp, 1976).

The unknown author occupied 11 pages showing "Practice" examples, with each page featuring a different "magic" rule. Thus, for example, the rule for "Case 4th" was:

> When the price is between one and two shillings, find the value of the quantity at 1 shilling per yard, &c, which value being divided by those even parts which the pence are of a shilling and the quotient or quotients arising therefrom added thereto the sum will be the answer.

Seven examples were given, the first of which was "$758\frac{1}{2}$ at 1*s* and 9*d* per yard."

In the traditional *abbaco* curriculum sequence, the topic "Practice" was placed well before the topic "Vulgar" (or Common) Fractions. In fact, in the North American cyphering books in the Ellerton-Clements collection that were prepared before 1792, most students who dealt with the topic "Practice" never proceeded to a formal study of vulgar fractions. Nevertheless, the aliquot parts of a shilling, which were essentially unit fractions of a shilling, were identified, and these were associated with divisions, as were aliquot parts of a pound, and of a yard, etc. Students also learned that a farthing might profitably be written as $\frac{1}{4}$, and a halfpenny as $\frac{1}{2}$. If, in later life, students faced real-life situations in which knowledge of a rule of Practice might have been helpful, they would then have needed to work out which of the numerous rules of Practice fitted the situation.

David Townsend's (1773) Introduction to "Fellowship": The Arithmetic of Partnerships

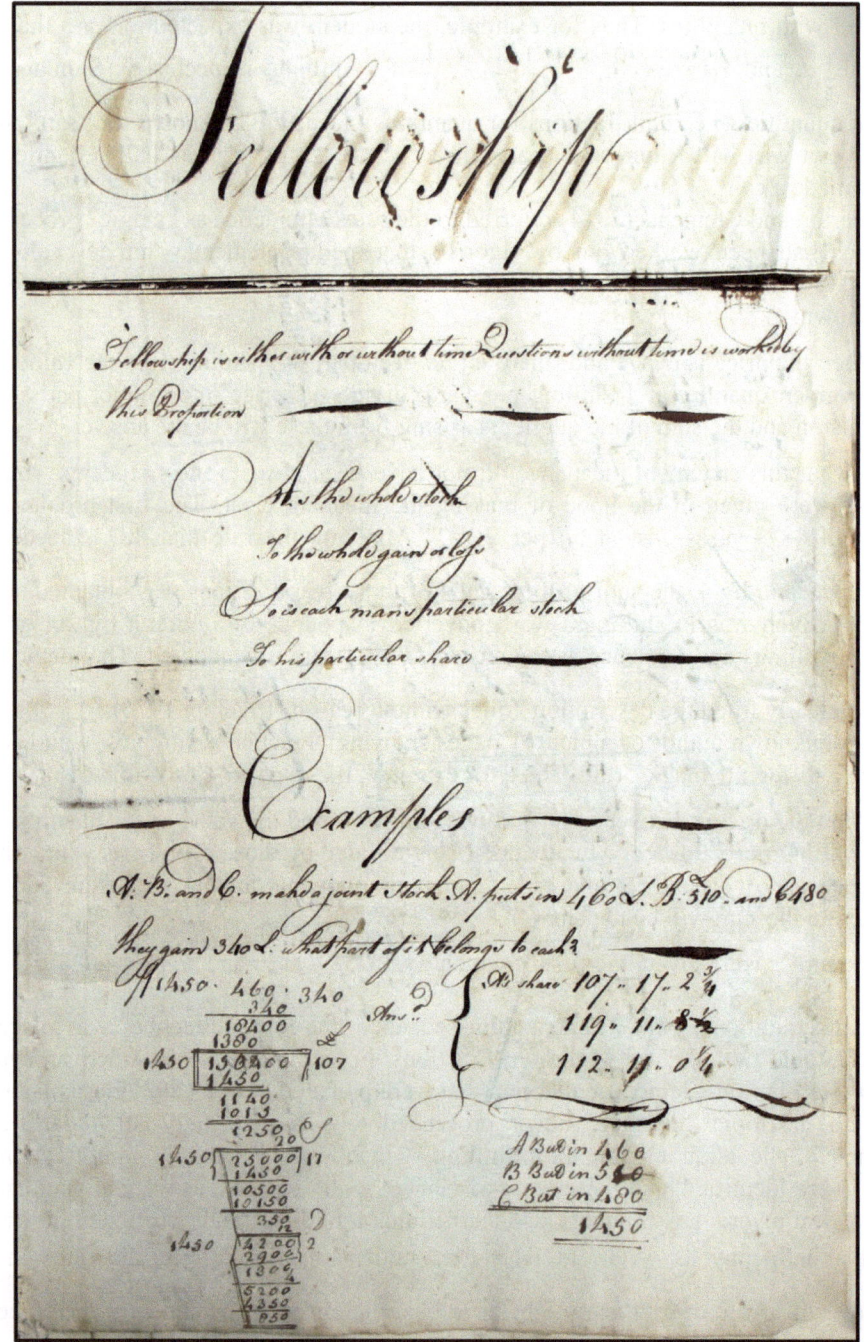

Figure 2.9. David Townsend's (1773) introduction to "fellowship."

Fellowship: The Arithmetic of Partnerships

Most eighteenth-century North American students who prepared cyphering books did not go beyond the rules of three (Ellerton & Clements, 2012). Of those who did, the topic of "Fellowship" was one of the most commonly studied higher-level aspects of arithmetic. There were two different types of fellowship—single fellowship, which applied when the main parties in a partnership contributed for an equal amount of time, and "double fellowship" (often called "fellowship with time") when parties were involved in an investment for different lengths of time. Nicolas Pike (1788) offered the following "rhymes" to describe the two forms of partnership:

Single Fellowship: "As the whole stock is to the whole gain or loss, so is each man's particular stock to his particular share of the gain or loss." (p. 155)

Double Fellowship: Multiply each man's stock, or share, by the time it was continued in trade, Then "As the whole sum of the products is to the whole gain or loss, so is each man's particular product to his particular share of the gain or loss." (p. 158)

Thus, for example, if in a joint venture Person A contributed 1000 pounds for a period of two years, and Person B contributed 1500 pounds for 18 months, and after the two years an overall loss of 1200 pounds was recorded, then the amount each person was responsible for contributing to the loss could be calculated as follows:

Person A: A calculation from the direct-rule-of-three situation would be:

Person A's contribution to the loss should be based on the rule-of-three statement:

4250: 1200 :: 2000 : $x,$ where Person A's contribution to the loss is £x .

Thus, Person A's contribution to the loss should be £$\dfrac{1200 \times 2000}{4250}$, or approximately

£564-14s. Hence, Person B's contribution should be approximately £635-6s.

Rationales for rules of fellowship rarely appeared in textbooks or cyphering books.

The verse which David Townsend (1773) included for "Fellowship" (shown in Figure 2.9) was essentially the same as the first of the fellowship rhymes given by Nicolas Pike (1788). David's solution to his joint stock problem was based on the rule-of-three statements:

For A:　1450 : 340 :: 460 : x, where A's profit would be £x;

For B:　1450 : 340 :: 510 : y where B's profit would be £y; and

For C:　1450 : 340 :: 480 : z, where C's profit would be £z.

Note that at the bottom right corner of Figure 2.9, a check might profitably have been shown—the sum of the three original gains should have been £1450 . In fact, the sum of the gains was one half-pence less than £1450—the difference being the result of rounding errors.

David Townsend prepared his cyphering book while living in Oyster Bay, Long Island. There can be little doubt that in a climate of investment, the skill of calculating gains or losses to be associated with a joint venture would have been useful, and much valued by some prospective employers. Other useful "higher-level" money topics, not discussed here, were loss and gain, foreign exchange, barter, simple and compound interest, discount, annuities, equation of payments, and false position (single and double).

Sally Halsey's (c. 1767) Entry on Alligation Alternate

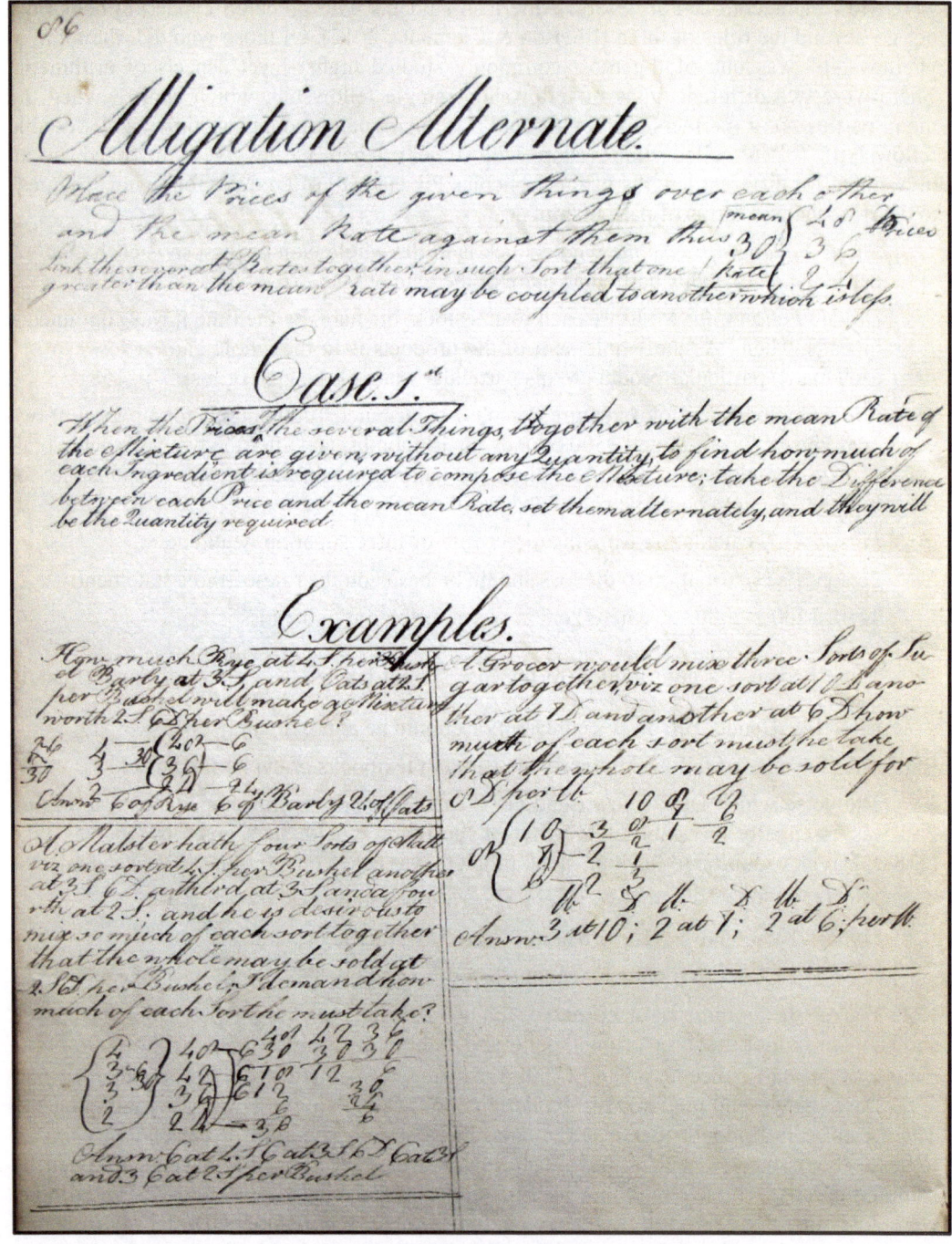

Figure 2.10. Sally Halsey's (c. 1767) introduction to "alligation alternate."

Did Cyphering Books Really Serve as Reference Books Later in Students' Lives?

Ellerton and Clements (2012, 2014) have provided data indicating that between 15 and 20% of all North American cyphering books were prepared by females. Sally Halsey (c. 1767), probably of New Jersey, was one of only a few eighteenth-century females who studied topics regarded as being beyond the rules of three. Alligation, which was concerned with mixing quantities, was one such topic, and in her cyphering book Sally Halsey included rules for "Alligation Medial," "Alligation Alternate," and "Alligation Partial."

Sally stated the following rule for "Alligation Alternate" (see Figure 2.10):

> When the prices of the several things, together with the mean rate of the mixture, are given without any quantity, to find how much of each ingredient is required to compose the mixture; take the difference between each price and the mean rate, set them alternately, and they will be the quantity required.

This rule would have been incomprehensible without examples, and the first example Sally considered was: "How much rye at 4/- per bushel, barley at 3/-, and oats at 2/- per bushel will make a mixture at 2s 6d per bushel? Sally's solution to this problem occupied just four short lines. The 2/- , 2/6, 3/- and 4/- were converted to 24, 30, 36 and 48 pence, respectively. Then numbers which were above the required mean (30) were linked with numbers below the mean (Sally linked 48 with 24, 36 with 24, and 24 with both 36 and 48). The prices (in pence per bushel) were listed vertically as 48, 36, and 24 and the differences between their linked numbers and 30 pence were listed opposite the three numbers (i.e., 6 was placed next to 48 because the difference between 24 and 30 was 6; another 6 was placed next to 36 because the difference 36 and 30 was 6; and 24 was placed next to the 24 because 18 + 6 = 24). Then the answer was stated as 6 bushels of rye, 6 bushels of barley, and 24 bushels of oats, and a check would show that that combination mixture would produce a mixture worth 2/6 per bushel (4 × 6 + 3 × 6 + 2 × 24 = 90, and 90 ÷ 36 = 2½).

This first "alligation alternate" problem for which Sally recorded her solution can be found in Thomas Dilworth's (1762, p. 84) *Arithmetic*. In fact, the problem was one of the simplest alligation alternate problems that could be devised. But the statement of the "alligation alternate" rule was so opaque that it is likely that Sally did not really understand what the rule meant. And, even if she did understand it, there is no indication in the text reproduced in Figure 2.10 that she understood the "reason for the rule." In her working she does not show how the 24 (on the right, opposite the 24 on the left) was obtained. Furthermore, there is no evidence that Sally was aware of, or considered, mathematically interesting questions like "was Sally's solution to the mixing problem unique?"

The *reason* for the rule that Sally used is not easy to articulate, and only a few of the small proportion of students who studied arithmetic as far as alligation alternate would have grasped it. From an eighteenth-century education perspective, that did not matter much— what *did* matter was whether Sally would have been able to solve a real-life version of the problem if, later in life, she found herself working for a merchant who mixed grains. Data are not available on whether students who recorded solutions to alligation problems (and other higher-level arithmetic problems) subsequently used their cyphering books as reference books to help them solve problems. There was an expectation that that should have been the case, and the issue would make an interesting theme for further research. Young adults, and apprentices, working on wharves at a major port like Salem, in Massachusetts, for example, may well have been expected to prepare mixtures like those indicated in Figure 2.10.

Cornelius Houghtaling (1775) on "Substraction" of Vulgar Fractions

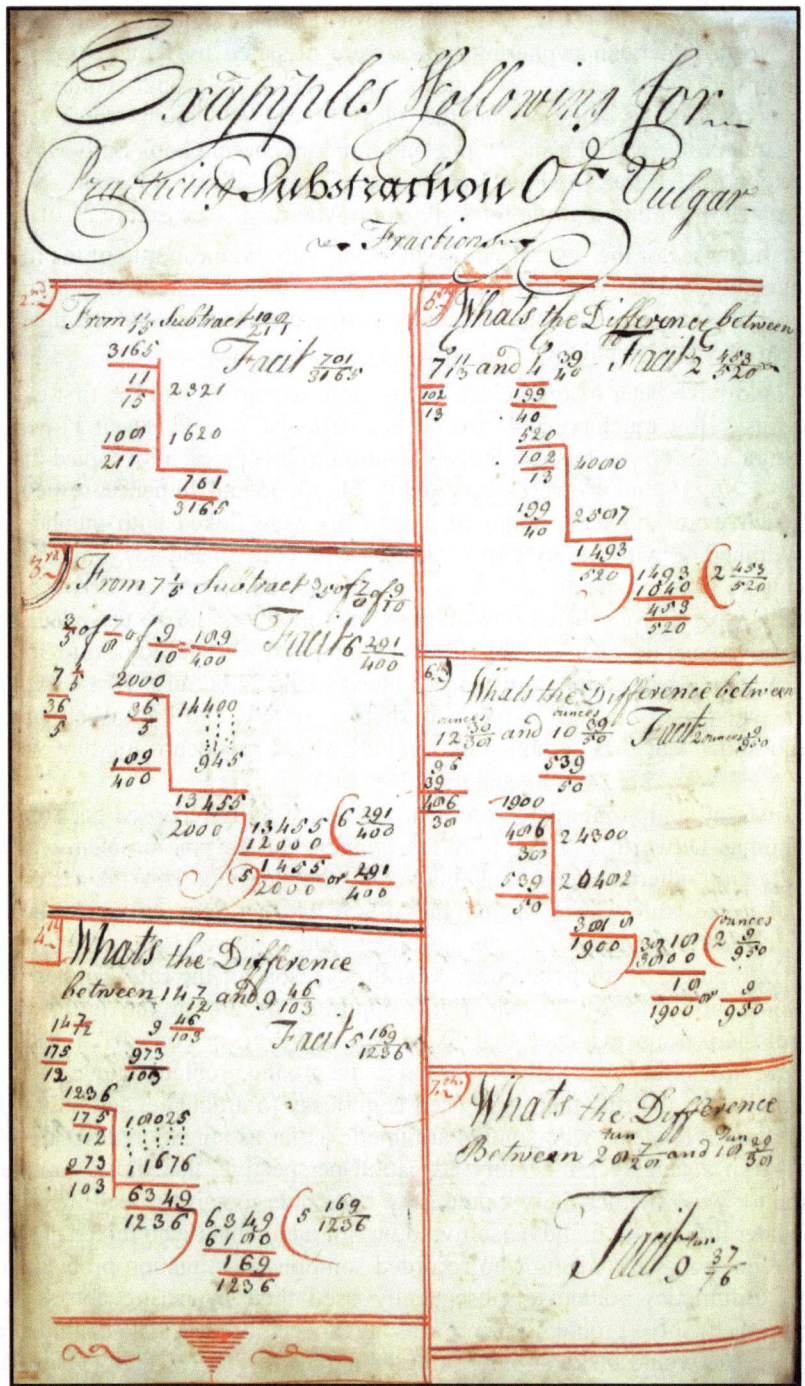

Figure 2.11. Cornelius Houghtaling (1775) subtracts vulgar fractions.

Cornelius Houghtaling's Rules for Subtracting with Vulgar Fractions

In June 1775 Cornelius Houghtaling, an 18-year-old from a Dutch-background family living in New Paltz, began to prepare a cyphering book. It was unusual for someone to begin a cyphering book in summer when most young men in and around New Paltz would have been expected to be out in the fields, making ready for the harvest. But in June 1775 the Revolutionary War with Great Britain was already underway, and around New Paltz these early Revolutionary years were full of tension and fighting (Forsyth, 1913). Young Cornelius had recently enlisted in the American army, and for some reason he decided that he should prepare a cyphering book. Genealogical research has revealed that Cornelius was from a well-to-do, slave-owning family, and before he began cyphering he secured an expensive full-leather large-sized, cyphering book with over 200 blank pages, each with dimensions 12 inches by 7 inches.

Why did Cornelius begin to cypher at such an advanced age? A possible answer to the question would be that he wished to earn money as a teacher, and he felt that he needed an attractive cyphering book to obtain such a position; he would also be able to use it if and when he did get a position. Another possibility is that, after quickly enrolling in the Revolutionary Army, Cornelius decided to prepare a beautiful cyphering when he was not involved in army duty. In addition, there was the distinct possibility that if he had a beautiful cyphering book, one that went further than the most elementary topics, he would be admitted to a prestigious college—such as the College of New Jersy in Princeton—after the War. For whatever reasons, we know that Cornelius prepared a 250-page manuscript, with each page carefully presenting two colors of ink.

Immediately before the calculations in Figure 2.11, Cornelius had considered the problem: "What is the sum of $\frac{6}{8}$ of a pound, $\frac{5}{7}$ of a shilling, $\frac{4}{5}$ of a penny, and $\frac{2}{3}$ of a farthing?" Cornelius had already made quite a few calculations on the previous page, and his calculations led to his remarkable answer of "15 shillings and $9\frac{1}{2}\frac{294912}{1955360}$." He then moved on to "Substraction of Vulgar Fractions." Cornelius began with "Case 1st":

> Case 1st: When a simple fraction is to be deductad (sic.) from a simple fraction reduce the fractions to a common denominator—Then take the numerator of the subtrahand (sic.) from the other and place the remainder over the common denominator. And you have the difference sought.

Then followed an illustrative example: "From $\frac{3}{4}$ take $\frac{5}{12}$." In order to do this, $\frac{3}{4}$ was first written as $\frac{9}{12}$, and "$\frac{4}{12}$ or $\frac{1}{3}$" was obtained.

Of the 21 North American students whose cyphering-book data were taken into account for Table 2.2, Cornelius was one of only two to do more than a minimal amount on vulgar fractions. Given the approach to vulgar fractions that he followed—see Figure 2.11—one wonders whether, for Cornelius, vulgar fractions ever became anything more than rules that had to be followed. Table 2.2 also indicated that Cornelius included a few pages on decimal fractions. Of the cyphering books on which Table 2.2 is based, only three other students included entries on both vulgar and decimal fractions.

Mahala Gove (1788) Subtracts and Multiplies Decimal Fractions

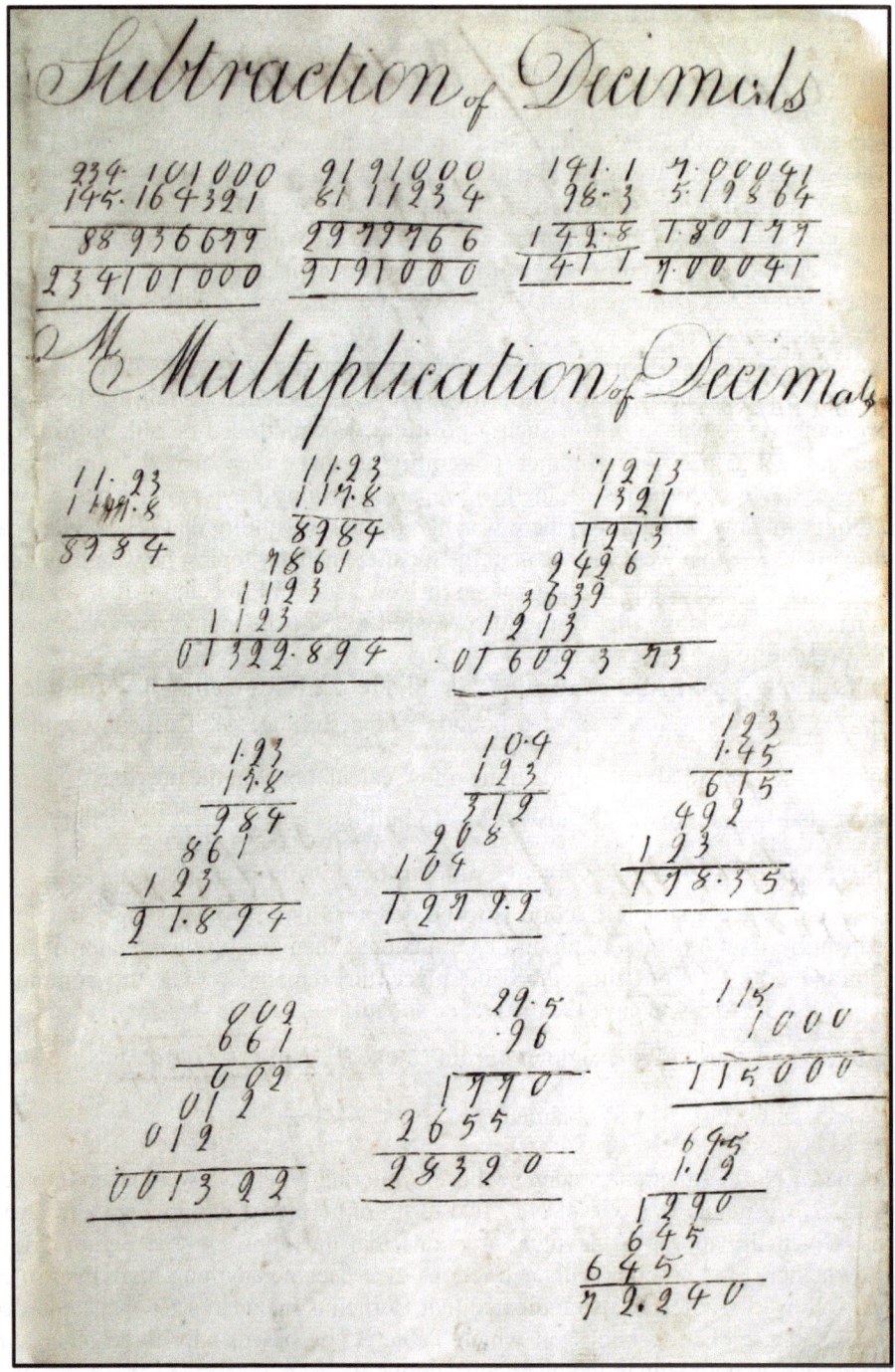

Figure 2.12. Operations on decimal fractions—by Mahala Gove (1788).

Mahala Gove (1788) and Decimal Fractions

Only 5 of the 20 cyphering books analyzed for Table 2.2 included entries on decimal fractions—those by Hulett Cornwell, Paul Cooledge, Cornelius Houghtaling, Loring Andrews, and Mahala Gove. Mahala's cyphering book, which was probably prepared in Hartford, Connecticut, comprised only 34 rag-paper pages, crudely sewn together. Five (that is to say, 14 percent) of those pages were concerned with decimal fractions. Figure 2.12 shows one of those pages. With each of the four tasks on subtraction of decimal fractions, a check, or "proof," was shown. On other pages there were tasks involving addition and multiplication of decimals, reduction of decimals, and the "rule of three decimals." At no stage in the cyphering book were there notes explaining what a decimal fraction was, or why one added, or subtracted, etc, decimal fractions in the ways which were illustrated. It is possible that the rather rough 34-page manuscript was merely a set of pages on which rough calculations were made before more careful entries were made in a fuller and neater cyphering book. It was an expectation of the cyphering tradition that all entries in cyphering books should be correct.

Mahala's working for three of the four subtraction tasks was unsatisfactory. With the first task (234.101000 − 145.164321) there is no decimal point placed in the "answer," and nor is there one placed in the "check." The answer given for the second task (9191000 − 8111234) is wrong, even if one assumes that the task really is 91.91000 − 81.11234. And the fact that the "check" was supposed to show that the calculations were correct when some of them were not, indicates that Mahala was not really carrying out the required check. The answer given for the third subtraction task (141.1 − 98.3) was also wrong, yet the check was intended to indicate that the calculations were correct. Mahala's incorrect answer for this third task provided evidence that she was not thinking about the meaning of subtraction—how could 142.8 be the answer if 98.3 were to be subtracted from 141.1?

Although the numerical calculations shown for the nine tasks under the heading "Multiplication of Decimals" were satisfactory, the placement of the decimal points in the answer was not always correct. With the third of the multiplication tasks, .1213 × 1321, the decimal point was not correctly placed for the answer. For the seventh task, it is not clear exactly which numbers were being multiplied—was the first number (the "multiplicand") .002, or was it 002, or was it something else? For the eighth task, 29.5 × .96 no decimal point was inserted in the "answer." And, with the ninth task, 115 × 1000, it is not clear what the multiplicand or the multiplier were, and no decimal point was given in the answer.

Inspection of other pages on decimals in Mahala's cyphering book revealed that irrespective of whether the operation was addition, subtraction, multiplication or division, she did not know where to place the decimal point in "answers." That uncertainty led to difficulties when she used decimal fractions when attempting to solve rule-of-three problems.

But at least Mahala was asked to study decimal fractions. The year was 1788, and Congress had agreed to establish a Federal currency which would be decimalized, and so it should have been clear that in the future a knowledge of decimal fractions would be helpful. So, there was some motivation for a teacher to decide to include decimal fractions in his (or her) implemented arithmetic curriculum. An important question that will be carefully considered in later chapters is whether, in the new nation, other teachers would regard it as an urgent matter to teach decimal fractions—because the nation had a decimalized currency. Or, would some teachers simply teach students how to express Federal currency in decimal form, and then to operate on those decimal forms by adding, or subtracting or multiplying, or dividing, or by carrying out formal methods of "reduction"?

The Cyphering Tradition's Grip on School Arithmetic

Figure 2.1, and Figures 2.3 through 2.12, and the comments accompanying those figures, provide a useful overview of the arithmetic curriculum studied by those students 10 years of age, or more than that, who prepared cyphering books in the mainly British colonies within the eastern part of North America during the seventeenth and eighteenth centuries. Most of the students who prepared cyphering books were male, and of those who did it was unusual to proceed to study topics "beyond the rule of three." Sally Halsey (a page of whose circa 1767 cyphering book is reproduced as Figure 2.10) provided a rare case of a female student who cyphered well beyond the rule of three.

The pages shown from the 21 cyphering books which were analyzed for the purpose of constructing Table 2.2 tended to have neat calligraphy and penmanship—although the quality of calligraphy and writing varied considerably between students (and even, sometimes, within a cyphering book). All of the entries were made with sharpened quills—feathers from birds—and usually the ink was home-made. Cyphering book pages typically featured the "Introduction-Rule-Case-Example-Exercise" (IRCEE) and the "Problem-Calculation-Answer" (PCA) genres identified by Ellerton and Clements (2012).

In most of the cyphering books there was a strong emphasis on money calculations, and the application of such calculations to topics like compound operations, reduction, loss and gain, practice, and the direct and inverse rules of three. Those who went beyond the rule of three would be likely to pay attention to more "advanced topics," like the double rule of three, barter, currency exchange, discount, simple and compound interest, annuities, equation of payments, fellowship, and false position. Such topics all had a strong money component to them, and derived from a European tradition designed to assist merchants to maximize profits in trading ventures (Cohen, 1982; Ellerton & Clements, 2012; Swetz, 1987; Van Egmond, 1980). Entries in Table 2.2 revealed that there was also a strong emphasis on weights and measures, all of which were not decimalized.

The most important aspect of this chapter came from the analyses summarized in Tables 2.1 and 2.2. There can be little doubt that in the North American British colonies the difference between author-intended curricula, as represented by the presence and balance of chapter themes in commercially-published arithmetic textbooks, on the one hand, and the presence and balance of themes in the student-prepared cyphering books, on the other, was considerable. That was particularly true with respect to vulgar fractions and decimal fractions. In 1775 almost two centuries had passed since Viète's (1579) and Stevin's (1585) pioneering work with decimal fractions, yet decimal fractions were not often studied by school students. With respect to decimal fractions, author-intended curricula differed from teacher-implemented curricula.

The fact that, in 1775, relatively few students studied decimal fractions—and the likelihood that most of those who studied them did not really understand them and did not learn to apply them accurately—provides the backdrop to one of the most interesting questions in relation to the history of school mathematics in North America: Would a government decision to introduce a Federal, decimalized form of currency, help to precipitate the planning of intended arithmetic curricula, and the adoption in schools of implemented arithmetic curricula, in which decimals fractions featured more strongly?

Of course, a similar question could have arisen if, sometime soon after 1776, the fledgling United States of America had decided to introduce a national system of weights and measures that was also decimalized. Educational issues associated with decisions by the U.S. Congress concerned with coinage weights and measures will be discussed in later chapters in this book.

References

[Note that handwritten cyphering books, to which reference has been made in the text, are not included in this, or in later reference lists in this book. Unless otherwise indicated, these cyphering books are part of the Ellerton-Clements collection.]

Bochner, S. (1966). *The role of mathematics in the rise of science.* Princeton, NJ: Princeton University Press.

Briggs, H. (1617). *Logarithmorum chilias prima.* London, UK: Author.

Burton, W. (1833). *The district school as it was, by one who went to it.* Boston, MA: Carter, Hendee and Company.

Cocker, E. (1677). *Cocker's arithmetick: Being a plain and familiar method suitable to the meanest capacity for the full understanding of that incomparable art, as it is now taught by the ablest school-masters in city and country.* London, UK: John Hawkins.

Cocker, E. (1685). *Cocker's decimal arithmetick, …* London, UK: J. Richardson.

Cohen, P. C. (1982). *A calculating people: The spread of numeracy in early America.* Chicago, IL: University of Chicago Press.

Denniss, J. (2012). *Figuring it out: Children's arithmetical manuscripts 1680–1880.* Oxford, UK: Huxley Scientific Press.

Devlin, K. (2011). *The man of numbers: Fibonacci's arithmetic revolution.* New York, NY: Walker & Company.

Dilworth, T. (1762). *The schoolmaster's assistant: Being a compendium of arithmetic both practical and theoretical* (11th ed.). London, UK: Henry Kent.

Ellerton, N. F., & Clements, M. A. (2012). *Rewriting the history of school mathematics in North America 1607–1861.* New York, NY: Springer.

Ellerton, N. F., & Clements, M. A. (2014). *Abraham Lincoln's cyphering book and ten other extraordinary cyphering books.* New York, NY: Springer.

Forsyth, M. I. (1913), The burning of Kingston, New York. *The Journal of American History, 7*(3), 1137–1145.

Gies, J., & Gies, F. (1969). *Leonard of Pisa and the new mathematics of the Middle Ages.* New York, NY: Thomas Y. Crowell.

Glaisher, J. W. L. (1873). On the introduction of the decimal point into arithmetic. *Report of the Meeting of the British Association for the Advancement of Science, 43,* 13–17.

Greenwood, I. (1729). *Arithmetic vulgar and decimal: with the application thereof, to a variety of cases in trade and commerce.* Boston, MA: S. Neeland & T. Green for T. Hancock.

Grendler, P. F. (1989). *Schooling in Renaissance Italy literacy and learning, 1300–1600.* Baltimore, MD: Johns Hopkins University Press.

Grew, T. (1758). *The description and use of the globes, celestial and terrestrial; with variety of examples for the learner's exercise.* Germantown, PA: Christopher Sower.

Hadden, R. W. (1994). *On the shoulders of merchants: Exchange and mathematical conception of nature.* New York, NY: SUNY Press.

Hill, J. (1772). *Arithmetick, both in the theory and practice, made plain and easy in the common and useful rules* … London, UK: W. Strahan.

Høyrup, J. (2008). The tortuous ways toward a new understanding of algebra in the Italian *Abbacus* school (14th –16th centuries). In O. Figueras, J. L Cortina, A. Alatorre, T. Rojano & S. Sepulveda (Eds.), *Proceedings of the joint meeting of PME 32 and PME-NA XXX* (Vol. 1, pp. 1–20). Morelia, Mexico.

Høyrup, J. (2014). Mathematics education in the European Middle Ages. In A. Karp & G. Schubring (Eds.), *Handbook on the history of mathematics education* (pp. 109–124). New York, NY: Springer.

Karpinski, L. C. (1980). *Bibliography of mathematical works printed in America through 1850* (2nd ed.). Ann Arbor, MI: University of Michigan Press.

Kersey, J. (1683). *Mr Wingate's arithmetick,* … (8th ed.). London, UK: J. Williams.

Lewin, C. G. (1970). An early book on compound interest—Richard Witt's arithmeticall questions. *Journal of the Institute of Actuaries, 96*(1), 121–132.

Lewin, C. G. (1981). Compound interest in the seventeenth century. *Journal of the Institute of Actuaries, 108*(3), 423–442

Napier, J. (1614). *Mirifici logarithmorum canonis descriptio.* Edinburgh, UK: Andrew Hart.

Napier, J. (1619). *The wonderful canon of logarithms.* Edinburgh, UK: William Home Lizars, 1857 [English translation by Herschell Filipowski].

Pike, N. (1788). *The new complete system of arithmetic, composed for the use of the citizens of the United States.* Newbury-Port, MA: John Mycall.

Radford, L. (2003). On the epistemological limits of language: Mathematical knowledge and social practice during the renaissance. *Educational Studies in Mathematics, 52*(2), 123–150.

Skemp, R. (1976). Instrumental understanding and relational understanding. *Mathematics Teaching, 77,* 20–26.

Smith, D. E., & Karpinski, L. C. (1911). *The Hindu-Arabic numerals.* Boston, MA: Ginn and Company.

Stevin, S. (1585). *De Thiende.* Leyden, The Netherlands: The University of Leyden.

Swetz, F. (1987). *Capitalism and arithmetic: The new math of the 15th century.* La Salle, IL: Open Court.

Van Egmond, W. (1980). *Practical mathematics in the Italian Renaissance: A catalog of Italian abbacus manuscripts and printed books to 1600.* Firenze, Italy: Istituto E Museo di Storia Della Scienza.

Venema, P. (1730). *Arithmetica of cyffer-konst.* New York, NY: Zenger.

Viète, F. (1579). Canon mathematicus seu ad triangula cum appendicibus. Paris, France: Jean Mettayer.

Wardhaugh, B. (2012). *Poor Robin's prophecies: A curious almanac, and the everyday mathematics of Georgian Britain.* Oxford, UK: Oxford University Press.

Wickersham, J. P. (1886). *A history of education in Pennsylvania.* Lancaster, PA: Inquirer Publishing Company.

Wingate, E. (1624). '*L'usage de la règle de proportion en arithmétique.* Paris, France: Author.

Wingate, E. (1630). *Of natural and artificiall arithmetique.* London, UK: author.

Workman, B. (1789). *The American accountant; or schoolmaster's new assistant.* Philadelphia, PA: John McCulloch.

Chapter 3
Thomas Jefferson and an Arithmetic for the People

"When I was young, mathematics was the passion of my life. The same passion has returned upon me [at age 69], but with unequal powers. Processes which I then read off with the facility of common discourse, now cost me labor, time, and slow investigation."

Thomas Jefferson to William Duane, October 12, 1812.

"I want to tell you how welcome you are to the White House. I think this is the most extraordinary collection of talent, of human knowledge, that has ever been gathered together at the White House, with the possible exception of when Thomas Jefferson dined alone.

Someone once said that Thomas Jefferson was a gentleman of 32 who could calculate an eclipse, survey an estate, tie an artery, plan an edifice, try a cause, break a horse, and dance the minuet."

Remarks of President John F. Kennedy at a Dinner honoring Nobel Prize Winners of the Western Hemisphere, held at the White House, April 29, 1962.

Abstract: This chapter focuses on U.S. currency issues, and related school mathematics practices during the period 1775–1792. As a result of post-war agreements spelled out by the Treaty of Paris (1783), the fledgling U.S. Congress needed to establish an official currency for the new nation. At that time, the most powerful financial figure in the nation was Robert Morris, Superintendent of Finance between 1781 and 1784, but aspects of his proposal for a new currency system were problematized by the young, and influential Thomas Jefferson, already a former Governor of Virginia, and famous for having drafted the Declaration of Independence. Like Morris, Jefferson proposed a decimal-based system of coinage, but the units for Morris's and Jefferson's systems were different. It was Jefferson's system which prevailed, and the most startling thing about his success on this matter was that his fundamental argument belonged to the realm of mathematics education—a combination of mathematics and education. Jefferson argued that his system would assist all U.S. citizens to achieve a better grasp of basic arithmetic than ever before, and that that would make it easier for them to survive with dignity. This chapter summarizes Jefferson's main arguments for the introduction of his version of decimal currency, and why he thought educational issues were so important.

Keywords: *Abbaco* curriculum; Arithmetic in U.S. schools; Decimal coinage; Decimal operations; History of the dollar; Notes on the establishment of a money unit; Robert Morris; Thomas Jefferson

Pre-War and War Currencies, Proclamation Money, and the Currency Crisis 1776–1792

The Declaration of Independence, July 4, 1776, formally marked the beginning of the United States of America but it was not until after the 1783 Treaty of Paris that the United States of America was internationally recognized as an independent nation. Before the War, the issuing of colonial paper money had been regulated by British Parliament Currency Acts of 1751, 1764, and 1773 (Greene & Jellison, 1961; Kleeberg, 1992; Ogg, 1927; Reid, 1991; Sosin, 1964), but between 1775 and 1783 the colonies made their own currency decisions. The Continental Congress had printed so much paper money that its value plunged until the expression "not worth a continental" became widely used at home and abroad (Lossing, 1851).

At the conclusion of the War, Congress faced a serious and urgent cash-generation and cash-flow problem—yet, until appropriate legislation was passed it could not directly levy taxes at a national level. As George Washington succinctly stated, the main problem with the new government was "no money" (Maier, 2010, p. 11). The Continental Congress had the power to print money, and did so; but, by 1786, the currency was virtually worthless. And, if Congress borrowed money, it would not have been able to pay it back, for the states were not paying all their U.S. taxes, and some states, like Georgia, were not paying any of their taxes. Interest due on debt owed to foreign governments was not being paid (Maier, 2010).

Congress could, and did, demand that states make contributions to national coffers, but demanding was one thing and receiving another. Between 1781 and 1784 less than one-and-a-half million Spanish dollars were passed on to the Continental Congress by the states, despite Congress having asked the states for much more than that (Goodwin, 2003; Hepburn, 1915). In 1785, Alexander Hamilton alerted the nation to the seriousness of the national cash-flow problem, and pointed out that the State of New York had paid nothing in taxes to Congress for that year. Many of the other states did little better than New York. The U.S. Congress recognized that, in order to be in a position to create a viable national currency, it needed to sell off, as quickly as possible, the sole major national asset—land, and especially land between the Ohio and Mississippi rivers (Linklater, 2003). But before it could do that agreements had to be reached with indigenous Americans, the traditional owners of the land, and the land had to be surveyed.

Even after 1783 the states continued to use their former colonial currencies which employed the British nomenclature of pounds, shillings, pence and farthings, and came in the form of paper notes. But, the values of notes of corresponding denominations varied from state to state (Hepburn, 1915). Although in 1788 the United States Constitution denied individual states the right to coin and print money, the use of existing currency notes continued. The First Bank of the United States, chartered in 1791, and the Coinage Act of 1792, began the era of a national American currency, which was to be based on coinage and not paper bills. Even so, the language of pounds, shillings, pence and farthings continued to be used in all 13 states, and sterling calculations continued to be taught in U.S. schools throughout the period 1783–1860 (Seaman, 1902). Spanish dollars also remained legal tender in the United States until 1857. That was the background to Thomas Jefferson's May 1788 statement that paper money was "only the ghost of money, and not money itself" (Thomas Jefferson to Edward Carrington, May 27, 1788). Jefferson believed that the value of an actual currency artifact (like, for example, a coin) should be equal to what that artifact could be exchanged for in day-to-day transactions.

Figure 3.1 shows a copy of a Pennsylvania three-pence "note," issued by John Dunlap in 1777 (Newman, 1990). Figure 3.2 reproduces one side of a 50-dollar currency bill (entitling "the bearer to receive 50 Spanish milled dollars, or the value thereof in gold or silver"), issued at Philadelphia in 1778. Figure 3.3 depicts the two sides of a Spanish dollar. Spanish dollars were widely used in the United States after 1776. These silver coins were often called "pieces of eight" because they were worth eight Spanish *reales* (or "royals"). Remarkably, the North Americans often cut the silver dollars into eight smaller pieces or "bits," and a "quarter" became known as "two bits." This provided the genesis for the fact that until the 1990s the New York Stock Exchange counted stock values in eighths of a dollar. Thus, in 1783 the citizens of the new United States of America were familiar with the idea of linking currency with vulgar or common fractions (Allen, 2009).

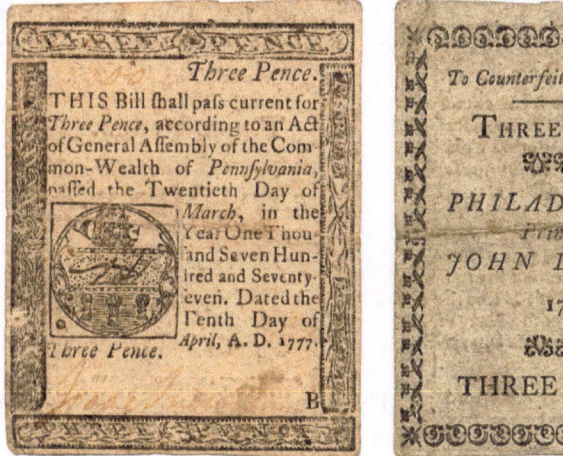

Figure 3.1. The two sides of a 3-pence Pennsylvania currency bill issued in 1777. Public domain images, USA.

Figure 3.2. A 50-dollar bill issued in 1778. Public domain image, USA.

Many purchases made before, during and after the Revolutionary War did not involve the use of paper money or coins, because *bartering* for goods was commonplace. In order to overcome the dilemma when someone wanted to obtain a commodity (for example, a bushel of oats) but did not have a form of currency the seller wanted to exchange for that commodity, the practice of using popular products (especially tobacco) as a kind of currency was adopted. Tobacco was a popular object of barter because it was easy to sell. Bartering often demanded the carrying out of tricky arithmetical calculations, and that is why the topic "barter" was part of the *abbaco* curriculum in some schools—although it was sufficiently difficult that relatively few students studied it formally.

Investors who traded with far-off countries (like France, or India, China, or the Spice Islands) employed "supercargoes" whose specific on-board task was to make sure that trading for goods was done profitably (Bartlett, 1933). One of the greatest-ever supercargoes, and someone who would become a fine mathematician and one of the world's finest navigators, Nathaniel Bowditch (1797), left a handwritten journal summarizing his trades, and the algorithms and principles he used when, as a supercargo, he bartered during a voyage from Salem, Massachusetts to Manila in 1796 and 1797.

Figure 3.3. Both sides, Spanish "milled" dollar used in the United States, 1776–1792. Public domain image, USA.

The practice of bartering was not new, for throughout much of the eighteenth century "rated commodities" like tobacco had been formally valued, in pounds, shillings and pence, and used for barter. The concept of "proclamation money" was formalized by an Act of the British Parliament in 1704, and this enabled exchanges in tobacco, for example, to be accounted in pounds, shillings, and pence. Many colonists paid their taxes with proclamation money (Brock, 1975; Massey, 1976; McCusker, 1992).

Currency Issues and the Education Challenge, 1783–1785

In the United States during the period immediately after 1783 there was a strong national feeling that, from the outset, the nation's political structures, and its schools and colleges, should take full advantage of the opportunity to create a unique and model democracy. A major question which needed to be resolved, and resolved quickly, was: "What should be the system of currency, and the associated coinage, for the new nation?" The two principal figures in the debate on this issue became Robert Morris—who, since 1781 had held the post of Superintendent of Finance for the Continental Congress (Frost, 1846; Linklater, 2003; Rappleye, 2010)—and the former Governor of Virginia, Thomas Jefferson (see Figures 3.4 and 3.5).

Morris, a Philadelphia merchant, land speculator, and former slave owner who had a large reputation for economic wisdom, had shown leadership with respect to financial matters during the War (Frost, 1846). But Jefferson, as the framer of the Declaration of Independence, was also widely respected. Remarkably, both Morris and Jefferson favored decimal systems of currency. At that time the only decimalized currencies in the world, and they were only partly decimalized, were in Russia and Japan. In 1704 in Russia Peter the Great had created a *ruble* which was equal to 100 *kopecks* (Fenzi, 1905); and in Japan—

where silver money was basically money by weight—1000 *mommes* was equal to 1 *kan* (Nishikawa, 1987).

The fundamental unit within Morris's proposed system would have had much less value than the Spanish dollar, which Jefferson put forward as the basis for the fundamental unit in his system. Jefferson proposed that decimalization should be formulated around a dollar roughly equal in value to the Spanish dollar, which had been much used in the United States during the period 1775–1784.

Figure 3.4. Thomas Jefferson (1743–1826)—Portrait by Rembrandt Peale (1800), held by the White House Historical Association (White House Collection). Public domain image.

Figure 3.5. Robert Morris (1734–1806).
Public domain image.

Robert Morris's and Thomas Jefferson's Currency Proposals

On July 6, 1785, the U.S. Congress resolved that "the money unit of the United States of America be one dollar," and that "the several pieces shall increase in decimal ratio." The unit would be a silver dollar—which would have a value about the same as the old Spanish dollar—and there would be a gold coin of 10 dollars, a silver coin worth a tenth of a dollar, and a copper coin worth one-hundredth of a dollar. There would also be a name to represent one-thousandth of a dollar. On August 8, 1786, Congress passed a resolution fixing the fineness of gold and silver coins at eleven-twelfth, the dollar to contain 375.64 grains of pure silver. It provided for a mill to be 1000th of a dollar, with coins that would be minted to represent half cents, double dimes (20 cents), half dollars (50 cents), dollars of silver, and five dollars and ten dollars of gold. On September 20, 1786, Congress passed an ordinance establishing a "mint" at which the coins would be made (Goodwin, 2003).

It is not appropriate here to provide details of the difficulties experienced by the fledgling U.S. government in establishing its decimal monetary system (see Hepburn, 1915 for such details). It suffices to say that the national mint, whose task was to produce the new coins, was not formally established until 1792, and no U.S. coins appeared until 1793. There can be no doubt, though, that Thomas Jefferson won a *personal* victory over Robert Morris, for the decimal system of currency which was established was basically the one that he had proposed (Jefferson, 1785; Linklater, 2003). Jefferson's aim was to create a new monetary system which would be "so simple that any farmer could do his own accounting" (Jefferson, 1785, p. 298).

The new monetary system was a *national* undertaking and from that perspective it is interesting that its chief proponent, Thomas Jefferson, was someone who was well-known for his determination to ensure that state rights were retained within the new Union. Jefferson seemed to have an ability to stand above state rights and jealousies when the situation absolutely demanded it. And the adoption of a national decimalized currency was one such case. The decision to adopt a decimal-based currency announced to U.S. citizens, and to the world, that the fledgling United States of America would not merely follow in the footsteps of older nations. From an *international* perspective, the U.S. decision on coinage meant that governments everywhere were forced to consider whether a decimal system provided the best model for dealing with financial matters within and between nations.

Thomas Jefferson regarded his successful campaign to establish the official national currency as a decimal currency as one of his greatest political achievements. Julian Boyd's (1953) account of Jefferson's efforts to gain preference for his decimal system over an alternative decimal system put forward by Robert Morris, who had exerted much influence on all financial matters during the closing years of the Revolution, brings out the way Jefferson was prepared to argue mathematically in order to win a case. To appreciate that point it will be useful first to look at Morris's proposal for a national currency.

Robert Morris's and Gouverneur Morris's Currency Proposal 1782–1783

In January 1782 Congress resolved to determine what currency arrangements should be made for the new nation, It began by summarizing the denominations of money currently being used in the states, and by providing a summary of the relative values of the foreign coins in common use in the United States. The Financier, Robert Morris was instructed to prepare a report on the coins in circulation within the United States, and Morris and his assistant Gouverneur Morris—the two Morrises were not blood-related—quickly prepared a lengthy handwritten report, which was forwarded to the President of Congress on January 15, 1782. The report, which was reproduced in full by Boyd (1953, pp. 160–169), would prove to be one of far-reaching importance. Among other things, it stated:

> In order that a coin may be perfectly intelligible to the whole people it must have some affinity to the former currency. This therefore will be requisite in the present case. the purposes of commerce require that the lowest divisible point of money or what is more properly called the money unit should be very small because by that means price can be brought in the smallest things to bear a proportion to the value. and although it is not absolutely necessary yet it is very desirable that money should be increased in a decimal ratio because by that means all calculations of interest exchange insurance and the like are rendered much more simple and accurate and of course more within the power of the great mass of people. Wherever such things require much labor time and reflection the greater number who do not know are made the dupes of the lesser number who do. (Quoted in Boyd, 1953, p. 166)

With these words the Morrises bravely recommended that the nation consider adopting a form of decimal currency. Since the only nations in the world at the time to have decimalized forms of currency were Russia and Japan, the Morrises' proposal was creative. Notice that there was an educational component to the proposal—establishing a decimal currency would make it easier for calculations associated with interest, insurance, and exchange to be carried out, so that they would come within "the power of the great mass of the people."

The Morrises also argued that the Spanish Dollar, which was in common use in North America, should be carefully considered when the new currency was formed, but they did not propose that the dollar be the standard unit. Rather, the Morrises argued that the unit should be much smaller than that. They pointed out that in Georgia the dollar was equal to 5 shillings, in North Carolina and in New York it equaled 8 shillings, and in Virginia and the four Eastern States it equaled 6 shillings. In South Carolina it equaled 32 shillings and 6 pence. Thus:

> The money unit of a new coin to agree without a fraction with all these different values of a dollar except the last will be the fourteen hundred and fortieth part of a dollar equal to the sixteen hundredth part of a crown. Of these units, 24 will be a penny of Georgia, 15 will be a penny of North Carolina, or New York, 20 will be a penny of Virginia and in the four Eastern States, 16 will be a penny of all the other States except South Carolina and 48 will be 13 pence of South Carolina.

So, the Morrises' unit would be $\frac{1}{1440}$ th part of a Dollar, and sums of money would be expressed as whole-number multiples of that unit.

Gouverneur Morris carefully explained where the 1440 came from in a letter to William Hemsley (dated April 30, 1783, and reproduced in Boyd, 1953, pp. 169–173). He wrote:

> The denomination of our coin must be found by an arithmetical operation which may be performed as follows. A dollar has four different nominal values in America, exclusive of the sterling rate adopted by S. Carolina.
>
> 1st   in Georgia where a Dollar is current for 60 pence
> 2dly   Virginia and the Eastern states   72
> 3dly   Jersey, Pennsylvania and Maryland   90
> 4thly   New York and N. Carolina   96
>
> A penny therefore of these several currencies is as follows:
>
> 1st   a penny Georgia money $\frac{1}{60}$ of a dollar
>
> 2.   a penny lawful $\frac{1}{72}$
>
> 3.   a penny proclamation $\frac{1}{90}$
>
> 4.   a penny New York $\frac{1}{96}$
>
> (Quoted in Boyd, 1953, p. 173)

Gouverneur Morris then declared that "the common denominator" was 1440.

Much more could be written, here, about the ingenious proposal put forward by the Morrises, but since it was not adopted it is sufficient to say that it was based on calculations with common fractions, and since most students in schools never got to study common fractions it would, at least initially, have generated confusion if it had been adopted. The Morrises probably recognized that that could have been the case; however, if their proposal had been accepted then conversions between state "pounds, shillings and pence" amounts and the new currency would have been facilitated—except, of course, in South Carolina.

Although the Morrises claimed that their proposed system was based on the idea of a "decimal ratio," it was only decimalized in the sense that the place-value feature of Hindu-Arabic numerals would have enabled calculations on whole numbers to take place within a decimalized system. By taking the currency unit as 1/1440th of a dollar, most calculations would have been with whole numbers.

It is possible that Robert Morris had self-interest in mind with his proposal. There can be no doubt that he was working toward being appointed director of a national mint (Boyd, 1953; Notes on the Early History of the Mint, 1867), and as Superintendent of Finance he had employed, at public expense, a certain Benjamin Dudley to prepare the way for the establishment of the Mint. Apparently, Robert Morris had promised Dudley that if the Mint were to be established then he would be appointed to the Mint. Dudley had even made a silver coin, and other coins, that would have fitted the Morrises' proposed system, and on April 17, 1783, these proposed coins were submitted to Congress (Boyd, 1953). But instead of adopting the proposal put forward by the Morrises, Congress created a new committee of five members to consider currency and coinage matters.

There were suspicions that Robert Morris was acting hastily on the matter because he had acquired one and a half million acres of land in Western Virginia and he wanted to sell that land, quickly, on profitable terms (Linklater, 2003). Undoubtedly, the Morrises were attempting to establish structures whereby the new nation could become financially viable, but Robert Morris also had his eye on securing legislation that would be profitable for land speculators like himself.

It was around this time that Thomas Jefferson became involved—though he was not on the committee of five established by Congress. Julian Boyd's (1953) account, written as an "editorial note" (see pp. 150–160), draws attention to the intricate politicking within and outside Congress on the matter, and to Jefferson's leading role within the negotiations. Boyd (1953), who commented that "Jefferson himself regarded this [the achievement of a decimal currency] as one of his principal legislative achievements in 1784" (p. 151), had this to say about Jefferson's work on the currency issue in 1783 and 1784:

Few, if any, of his [i.e., Thomas Jefferson's] state papers present a better example of his practical statesmanship and legislative skill than these pages in which he reduced an unusually complex subject to a simplicity that made his argument against the Morrises' rival plan overwhelming. But brilliant clarity and simplicity of presentation, though they persuaded legislators to adopt the dollar as the American unit of currency and, for the first time in history, to apply the decimal system of reckoning to coinage, are not the only distinguishing features of this remarkable production of his pen: there is in it, to an unusual degree, many of the cardinal features of his philosophy—his attitude toward "the bulk of mankind," his concept of an evolving society, his love of order and system, his desire to cut new channels of trade toward the continent and away from England, his wish to see the intellectual as well as the man of affairs have influence ("This is a work proper to be committed to mathematicians as well as merchants"), his realistic acceptance of the strength of social habits, and his avoidance of ingenious solutions that disregarded such factors, his serene acceptance of the slow processes of history, ... (From Boyd, 1953, pp. 150–151)

Boyd was effusive about all the factors that Jefferson seemed to take into account when dealing with the currency/coinage issue in 1783 and 1784—but he failed to identify one aspect of Jefferson's thinking that was clearly in the forefront of Jefferson's mind at the time, and one which would profoundly affect the future of the United States of America. In putting forward his proposals on coinage, Jefferson clearly had in mind mathematics education and, in particular, the development of number sense within society at large. That issue will now be taken up in detail.

Thomas Jefferson's Educational Basis for a Decimalized System of Coinage

The Dollar Becomes the Unit of U.S. Currency

It was probably in February or March 1784 that Thomas Jefferson prepared a brief, undated, handwritten document which he titled "Some Thoughts on a Coinage" (the text is reproduced in Boyd, 1953, pp. 173–175). According to Julian Boyd (1953, p. 155) the text of this document was never published before Boyd himself included it in Volume 7 of *The Papers of Thomas Jefferson*. Commentators on Jefferson's contributions to the debate on currency have been captivated by Jefferson's longer "Notes on Coinage," which he prepared, sometime between March and May 1784—a printed version of that document appears in Boyd, 1953, pp. 175–188)—not long after he had prepared after "Some Thoughts on a Coinage." In a footnote, Boyd (1953, p. 175n), commented that "Some Thoughts on a Coinage" was erroneously placed with a July 1790 document that Jefferson would prepare on weights and measures. For whatever reasons, "Some Thoughts on a Coinage" has not been given the attention it deserves. Historically, it is of major importance, for reasons which will now be discussed.

In the first few lines of "Some Thoughts on a Coinage"—the text of which is given in Appendix *A* to this book—Jefferson noted that "the size of the unit" needed attention—subsequent events would make clear that Jefferson was dissatisfied with the size of the unit of currency proposed by the Morrises. He was particularly concerned with what he called the "division" of the unit, and "its accommodation to known coins" (Boyd, 1953, p. 173).

Then Jefferson proceeded to summarize his thoughts on a "transition from weights to measures" (p. 173). Even though "Some Thoughts" was merely a summary of what Jefferson was thinking, it is obvious that he had conceived of a system by which U.S. money and the measurement of length, area, volume, capacity, weight and time would be synthesized into one coordinated system. Jefferson had conceived of this plan, which was not dissimilar to the plan for weights and measures instituted in France after the French Revolution, *before* Jefferson left for France (in July 1784). Jefferson's "Some Thoughts" made clear that his unit for time would come from "a rod vibrating seconds," which needed to be "nearly $58\frac{1}{2}$ inches" (Boyd, 1953, p. 175). He wrote of a "geometrical mile" being 6086.1 feet or "1863.4 second pendulums" and an "American mile" being 6086.4 feet (Boyd, 1953, p. 174). Length measures would also be decimalized, so that 1 furlong would equal 608.64 feet, 1 chain would equal 60.864 feet, and 1 pace would equal 6.0864 feet.

Jefferson realized how revolutionary his idea was. In a letter he sent to Francis Hopkinson on May 3, 1784 he wrote: "In the scheme for disposing of the soil an happy opportunity occurs for introducing into general use the geometrical mile in such a manner as that it cannot possibly fail of forcing it's (*sic.*) on the people" (see, Thomas Jefferson to Francis Hopkinson, May 3, 1784, in Boyd, 1953, p. 205). Jefferson knew that the introduction of his new synthesis for measure was so radical that it needed to be "forced" on the people. In his letter to Hopkinson, Jefferson considered the possibility that some legislators might object because his idea was too much related to "astronomy and science," and predicted that conservatives would argue that it would be good to retain the 12-penny pound and the 12-inch foot "to preserve an athletic strength of calculation" (p. 205).

After he had prepared "Some Thoughts on a Coinage" in February or March 1784, Jefferson apparently decided that it would be unwise to attempt at that stage to persuade Congress to accept his radical synthesis of coinage and weights and measures, and instead he prepared a comprehensive statement on "Notes on Coinage" in which he said nothing about the weights and measures aspect of his earlier summary statement—except, of course, the weights of gold and silver in the coins he was proposing. In "Notes on Coinage" he proposed a decimal sequence of coins, with the dollar as the unit. His proposal differed from that put forward by the Morrises because his unit would be much larger than the unit the Morrises proposed. As a result of his larger unit, one would think of amounts of money that would be represented as, for example, $17.12 (although Jefferson did not propose the "$" symbol). $17.12 would represent 17 dollars and 12 hundredths of a dollar. In addition to a silver dollar, there would be a smaller silver coin for a tenth of a dollar, and there would be a copper coin for a hundredth of a dollar; and there would also be a golden coin for ten dollars. So, $17.12 would be the value of one gold piece, seven silver dollars, one silver "one-tenth of a dollar" and two copper "one-hundredths of a dollar." In July, 1784, soon after he had prepared his "Notes on Coinage," Jefferson left for Paris where he remained for five years as Minister Plenipotentiary to France. By the time he returned to the United States, his friend James Madison had managed to shepherd the coinage proposal through Congress, and in 1792 a national mint was established (see Boyd, 1953, pp. 150–204 for an account of the sequence of political events by which this occurred. See also Appendix *A* to this book).

Jefferson's Focus on Simplifying School Arithmetic

Of special interest for this present work is how the U.S. decision to decimalize currency presented a large challenge for those concerned with teaching citizens, especially children, how to think about and operate with amounts of money expressed as decimal fractions. There were no existing textbooks available to explain how children might be expected to apply the new decimal system in the shops and in the schools. Merchants would need to learn to think in eagles, dollars, cents and mills for local, interstate and international trade. Teachers in the United States had had no experience using, or teaching about, decimal currency, and many of them knew very little about decimal fractions (Ellerton & Clements, 2014).

In Chapter 2, evidence was presented from children's handwritten entries in cyphering books on how children learned to make money calculations *before* the advent of decimal currency. Basically, calculations were made on amounts of money represented in the sterling units of pounds, shillings, pence and farthings. Such calculations often demanded that children switch from base 4 (with farthings) to base 12 (with pence) to base 20 (with shillings). Then, with whole numbers of pounds, base 10 arithmetic was needed.

In his "Notes on Coinage" Jefferson argued persuasively that his decimal system of currency was what was needed for the new nation because, of all possible currency arrangements, it would be most easily understood, and used by the citizens of the new nation. Despite the pressure of having to deal with last-minute matters before he departed for France, Jefferson devoted much time to the preparation of his justification for decimal currency.

At no stage did Jefferson mention that only two nations in the world—Russia and Japan—had decimalized units in their currencies. His main argument in favor of a decimalized currency was that most people would find decimal calculations easy to do. After boldly asserting that "the most easy ratio of multiplication and division is that by ten" (quoted in Boyd, 1953, p. 176), he added:

> Every one remembers, that, when learning money-arithmetic, he used to be puzzled with adding the farthings, taking out the fours and carrying them on; adding the pence, taking out the twelves and carrying them on; adding the shillings, taking out the twenties and carrying them on; but when he came to the pounds, where he had only tens to carry forward, it was easy and free from error. The bulk of man-kind are school-boys through life. These little perplexities are always great to them. ... Those who have had occasion to convert the *livres, sols*, and *derniers* of the French; the *gilders, stivers, and frenings* of the Dutch; the pounds, shillings, pence and farthings of these several states, into each other, can judge how much they would have been aided, had their several subdivisions been in a decimal ratio. (Quoted in Boyd, 1953, p. 176)

Jefferson then offered model examples to show how decimal calculations might simplify money calculations, comparing the traditional "English" sterling ways of adding, subtracting, multiplying, and dividing money, with corresponding calculations that might be expected with his proposed system. He set out calculations for addition and subtraction that every person who had ever prepared a cyphering book would have recognized.

Jefferson's over-simplification of associated educational issues. The following is what Jefferson wrote for how equivalent amounts of money expressed in pounds, shillings, pence and farthings, on the one hand, and in dollars and cents, on the other, would be added and subtracted (from "Notes on Coinage," reproduced in Boyd, 1953, pp. 176–177):

	£	s.	d	qrs.	Dollars.			£	s.	d	qrs.		
		Dollars.											
Addition.	8	13	11	1-2	=	38.65	*Subtraction.*	8	13	11	1-2	=	38.65
	4	12	8	3-4	=	20.61		4	12	8	3-4	=	20.61
	13	6	8	1-4	=	59.26		4	1	2	3-4	=	18.04

Jefferson made it all look so easy. But what he did not say was that despite the fact that most people were used to dealing in a currency which used the nomenclature "dollars," the dollars were Spanish dollars, and Spanish dollars were not decimalized. It is possible that Jefferson was not aware that, despite the fact that chapters on decimal fractions appeared in most school arithmetic textbooks, without further education only a tiny percentage of North Americans—including persons likely to be teachers of mathematics—would have known, for example, what 38.65 Jeffersonian dollars meant. If Jefferson's scheme were to be accepted by Congress then a massive education program in the schools, and a large-scale public re-education initiative outside the schools, would be needed. On this point, see Appendix *B*.

After his addition and subtraction examples, Jefferson compared calculations for non-decimal and decimal currency under multiplication and division. He presented the following cases:

Multiplication by 8.					*Division by 8.*				
£	s.	d	qrs.	Dollars.	£	s.	d	qrs.	Dollars.
8	13	11	1-2 = 38.65		8	13	11	1-2 =	8⟨38.65
20				8	20				4.83
173				309.20	173				
12					12				
2087					2087				
4					4				
8350					8⟨ 8350				
8					4⟨ 1043				
4⟨66800					12⟨ 260 3-4				
12⟨16700					20⟨ 21 8 3-4				
20⟨1391 8					£1 18 3-4				
£69 11 8									

Jefferson asserted that "a bare inspection of the above operations will evince the labor that is occasioned by subdividing the unit into 20ths, 240ths, and 960ths, as the English do, and as we have done; and the ease of subdivision in a decimal ratio" (from "Notes on Coinage," in Boyd, 1953, p. 177).

Jefferson exaggerated the difficulty for multiplying and dividing non-decimal quantities. When dividing 8 pounds, 13 shillings, 11 pence and 2 farthings by 8 he converted the 8 pounds, 13 shillings, 11 pence and 2 farthings to 8350 farthings and then divided by 8, 4, 12, and 20. But our analyses of textbooks and cyphering books has indicated that most students would not have multiplied or divided 8 pounds, 13 shillings, 11 pence and 2 farthings by 8 using the method that Jefferson claimed was common.

The algorithm employed by most textbook authors and most students resulted in the following approach: "8 into 8 pounds goes 1; 8 into 13 goes 1 and 5 (shillings) left over; 8 into 71 goes 8 and 7 (pence) left over; 8 into 30 goes 3 and 6 left over. Therefore, not considering the fraction of a farthing left over, the answer is 1 pound, 1 shilling, 8 pence and 3 farthings." A similar procedure, avoiding conversion to 8350 farthings, could have been used with multiplication by 8. Using those procedures would have occupied far less space in the setting out than the algorithms which Jefferson claimed were the most commonly used.

To illustrate the point, consider the three examples of division of money shown in Figure 3.6, which are from a page that John McDuffee (1785) prepared at a school in Rochester, New Hampshire. John rather untidily, but nevertheless very succinctly, set out his responses to the following three tasks:

- Divide 36 pounds, 10 shillings and 6 pence equally among 5 men.
- Divide the Carges (*sic.*) of a country feast amounting to 468 pounds, 12 shillings and 4 pence equally among 12 stewards to see how much each steward should pay.
- Divide 57 pounds 9 shillings and 4 pence among 15 men.

Jefferson's claim that division of sums of money expressed in pounds, shillings and pence was likely to take far more "labour" than if an equivalent amount expressed in a decimal form is not supported by the John McDuffee's (1785) entries (shown in Figure 3.6). It is stressed, here, that McDuffee's approach was typical of how division of sterling money tasks was set out in cyphering book entries by most students studying division of money.

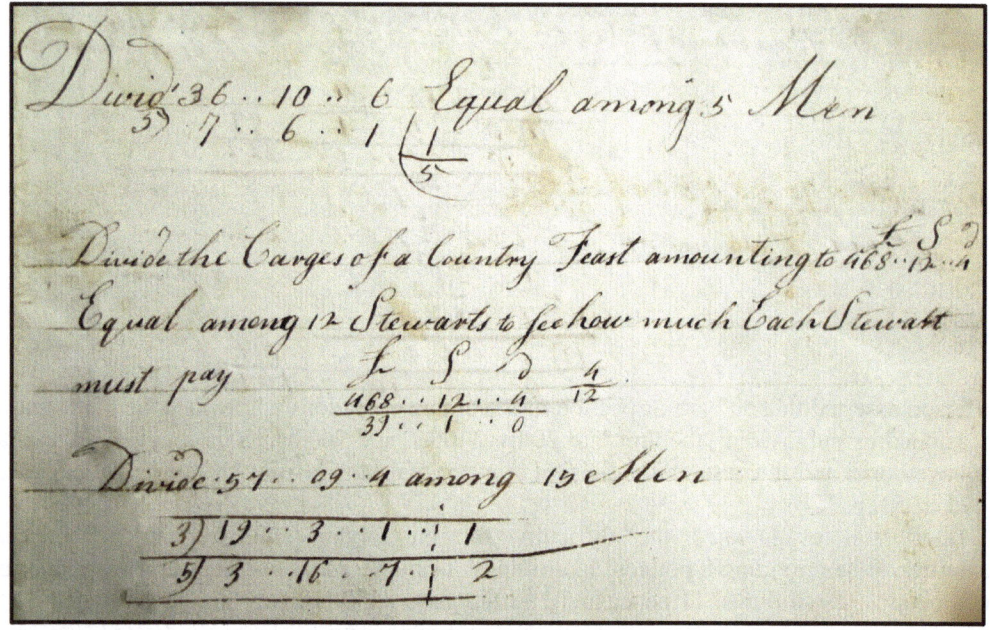

Figure 3.6. John McDuffee's (1785) setting out for dividing sterling currency.

Jefferson was confused between the method used for multiplication of money (expressed in sterling) and the method used for "reduction of money." In fact, the method he used to show that 8 pounds 13 shillings and $11\frac{1}{2}$ pence was equal to 8350 "quarters" was the method used by textbook authors and students for reduction tasks. That Jefferson was confused concerning the structuring and sequencing of school arithmetic was not surprising given that it was over 25 years since he had studied school arithmetic himself. Furthermore, his own privileged form of schooling had been quite different from that which was experienced by most U.S. children in the mid-1780s. In fact, in the 1780s only a tiny proportion of U.S. schoolchildren learned rules for operating with decimal fractions—that was because they never studied or worked with decimal fractions at school. Even those who did learn the rules were unlikely to know why those rules "worked." The students were happy merely to attempt to follow the stipulated rules.

Jefferson did not realize that if his scheme were accepted it would likely be many years before most adults would be able to perform multiplications and divisions using decimals. Jefferson seemed to think that if a system of decimal currency were to be introduced then money calculations would suddenly become very easy for most school children and adults. He concluded his display of what he thought was school arithmetic with the comment: "An

Englishmen to pay 8 pounds 13 shillings 11 pence and 2 farthings must find, by calculation, what combination of the coins of his country will pay this sum; but an American will have the same sum to pay thus expressed $38.65, will know by instruction only, that 3 golden pieces 8 units or dollars, 6 tenths, and 5 coppers, pay it precisely" (Jefferson, 1785, p. 300). But, it is hard to imagine that an English person would have had difficulty in deciding which coins would be needed to display 8 pounds 13 shillings 11 pence and 2 farthings.

Jefferson's Big Picture for Currency Reform

Five Features of Jefferson's Big Picture for Currency Reform

Jefferson had a passion for mathematics, and was very good at it (Fauvel, 1999). But, because he had had no direct experience in teaching it to schoolchildren there could be no guarantee that what he wrote about money calculations corresponded to what students in the schools would think or would enter into their cyphering books. Most likely, a similar statement would also have been true for other topics in school arithmetic. That was why in his "Notes on Coinage" Jefferson got the details wrong with respect to common methods for multiplying and dividing sums of summary expressed in sterling currency. Few, if any, of the members of Congress who read Jefferson's claims would have known enough about what ordinary people could and could not do by way of calculation, to be aware of his educational over-simplifications. He was keen to introduce a system which would make it possible for ordinary people to understand which monetary calculations were needed for everyday events, and how and why those calculations could be performed. He seemed to regard that as more important than enabling traders and business leaders to profit through monetary transactions which only they understood and which they controlled. Like Morris, Jefferson wanted to introduce a system which might help relieve the crippling national debt—he did not object to reasonable profits being made by land speculators, provided those profits were not the result of currency manipulations that only the wealthy could engineer through financial advisers.

But ease of calculation was not the only aspect of the big picture Jefferson envisaged for a new currency system. He has rightly been called a sage, a dreamer, someone who attempted to sketch and then seek to implement "big pictures." With the coinage issue it is possible to identify, especially but not only from "Some Thoughts on a Coinage" (reproduced in Boyd, 1953, pp. 173–175) and "Notes on Coinage" (reproduced in Boyd, 1953, pp. 175–185), the following five underlying principles that were important features within his big picture:

1. The fundamental unit of currency should be of suitable size—not too small, and not too large;
2. Because of the need to announce to the world that the United States of American was freeing itself of unsuitable vestiges from its colonial past, there should be a firm decision to drop the sterling nomenclatures of pounds, shillings, pence and farthings;
3. The monetary system should have a decimal basis;
4. Both mental and written calculations with the new monetary system should be sufficiently straightforward that "ordinary people" would be able to cope with them, readily; and
5. The new monetary system should be consistent with the idea of introducing a grander coordinated system of measurement which embraced money, time, length, area, volume, capacity, and perhaps even angle.

It will be helpful to comment briefly on each of those five underlying aspects.

1. The fundamental unit of currency should be of a suitable size. Jefferson made clear that he thought the purchasing power of a Spanish dollar—a currency widely used in the fledgling United States of America during the 1770s and 1780s was appropriate to be translated into the main unit in his proposed system of currency. Jefferson said he could not think of "a common measure of more *convenient size*" ("Notes on Coinage," Boyd, 1953, p. 175, Jefferson's emphasis retained). The unit would be appropriate for large-money purchases involving, say, "100, 1000, 10,000 dollars" (p. 175), while at the same time, using decimal fractions, it would be suitable for small everyday purchases which required only mental calculation. Jefferson wrote:

> The expediency of attending to the size of the Money Unit will be evident to any one who will consider how inconvenient it would be to a merchant, if, instead of the yard for measuring cloth, either the inch or the mile had been made the Unit of measure. (p. 175)

In "Notes on Coinage" Jefferson specifically criticized the Morrises' proposal on coinage because their fundamental unit would have been too small (p. 179). Historical perspective would seem to suggest that with respect to the size of unit, Jefferson was correct.

2. The need to be rid of colonial vestiges. Jefferson was worried that some citizens, unable to free themselves from colonial vestiges, might want "to hang the pound sterling, as a common badge, about all their necks" (p. 178). Despite the recent War, there was a real danger of that happening because various versions of sterling currency continued to be used in all 13 states. The people were familiar with the sterling terminologies, and Jefferson knew that it would be "difficult to familiarize them to a new coin with an old name" (p. 178). Wisely, Jefferson chose the name "dollar" for his new currency—the people had been using Spanish dollars for years, and since the new U.S. dollar would have approximately the same value as the Spanish dollar, the transition would be less difficult.

Even so, Jefferson realistically recognized that in any new scheme of currency, the remnants of the old colonial pounds, shillings and pence would be slow to disappear. From that perspective, he thought that his proposal for the introduction of an American dollar based on the Spanish dollar would please "the people" because they already knew what a Spanish dollar could buy, and because they already knew how to convert backwards and forwards between the dollar and the local system of pounds, shillings and pence. He wrote:

> They now have two units, which they use with equal facility, viz. the pound of their respective state, and the dollar. The first of these is peculiar to each state: the second, happily, common to all. In each state the people have an easy rule of converting the pound of their state into dollars, or dollars into pounds; and this is enough for them, without knowing how this may be done in every state of the Union. Some of them as live near enough the borders of their state to have dealings with their neighbors, learn also the rule of their neighbors; thus in Virginia and in the Eastern States, where the dollars is 6s, or 3-10 of a pound, to turn pounds into dollars, they multiply by 10 and divide by 3. To turn dollars into pounds, they multiply by 3 and divide by 10. Those in Virginia who live near to Carolina, where the dollar is 8s, or 4-10 of a pound, learn the operation of that state, which is a multiplication by 4, and division by 10, *et e converso*. Those who live near Maryland, where the dollar is 7s. 6d. or 3-8 of a pound, multiply by 3

and divide by 8, *et e converso*. All these operations are easy and have been found by experience, not too much for the arithmetic of the people when they have occasion to convert their old unit into dollars or the reverse. ("Note on Coinage," in Boyd, 1953, pp. 183–184)

Jefferson could see where, ultimately, he wanted the nation to go, so far as currency was concerned, but he was prepared for short-term compromises. Despite natural animosities in the United States of America toward Great Britain after the Revolutionary War (School Histories and International Animosities, 1903), local versions of the sterling currency's pounds, shillings and pence continued to be used across North America until the second half of the nineteenth century.

3. The need for a decimal basis for the new currency. Jefferson recognized that this could be the most contentious aspect of all. Only a handful of people in the United States would have known that decimal relationships among some coins existed in Russia and in Japan—so, they would have wondered why there was a need to depart from the pounds-shillings-pence-farthings relationships that they had grown to understand. Jefferson recognized that there could be another argument advanced by those who did not favor a decimal currency—specifically, it might be argued that having to deal with the 20-12-4 ratios associated with pounds, shillings and pence was a positive feature because it would not only provide the populace with practice in making important arithmetical calculations, but would also require those making the calculations to think about what they were doing. In his letter to Francis Hopkinson of May 3, 1784, Jefferson referred to this as a misguided belief that the 20-12-4 ratio helped to "preserve an athletic strength of calculation" (Thomas Jefferson to Francis Hopkinson, May 3, 1784, in Boyd, 1953, p. 205). He maintained that that was a foolish attitude, because this was "surely an age of innovation, and America the focus of it!" (p. 206—the exclamation mark was included by Jefferson in the original statement).

4. Currency considerations should be associated with straightforward calculations. Jefferson was clear that it was important in cases where there was a choice "between easy and difficult modes of operation, it is most rational to chuse (sic.) the easy" ("Notes on Coinage," Boyd, 1953, p. 176). It was the need for a system which would encourage straightforward calculations which led Jefferson to compare the ease of adding 4 pounds 12 shillings $8\frac{3}{4}$ pence to 8 pounds 13 shillings $11\frac{1}{2}$ pence with that of adding \$20.61 to \$38.65 (and with other calculational tasks involving subtraction, multiplication and division). Jefferson's attitude was guided by a perceived need to empower the people so far as currency calculations were concerned. In that sense, when dealing with currency issues Jefferson always had the mathematical education of the American people in the back of his mind.

5. The new coinage system should be consistent with the introduction of a grander decimalized scheme of weights and measures. Jefferson's "Some Thoughts on a Coinage" made clear that he regarded the coinage issue as intimately related to a wider issue concerned with the introduction of a national system of weights and measures—which was also to be decided on a national basis. Jefferson had developed a grand scheme by which time, length, area, volume, capacity, weight and money were to be connected through a decimalized system of measures. There were many in Congress who would be unlikely to support such a grand scheme. In 1782, the Financier, Robert Morris, had informed the President of Congress:

It is very fortunate for us that the weights and measures used throughout America are the same. Experience has shewn in other countries that the efforts of the legislator to change weights and measures although fully seconded by the enlightened part of the community have been so strongly opposed by the popular habits and prejudices that ages have elapsed without producing the desired effect. I repeat therefore that it is happy for us to have throughout the Union, the same ideas of a mile and an inch, a hogshead and a quart, a pound and an ounce. So far our commercial dealings are simplified and brought down to the level of every capacity. (Robert Morris to the President of Congress, January 15, 1782, reproduced in Boyd, 1953, pp. 160–161)

This statement was palpably false, for although each state had officially accepted troy, avoirdupois and apothecaries systems of weights that Queen Elizabeth had imposed on Great Britain in the sixteenth century, a hogshead in one state, or even territory, was likely to be different from a hogshead in another. The same was true of virtually every other named unit in the troy, avoirdupois and apothecaries systems (Linklater, 2003). Much confusion reigned, but that favored, for example, the rich merchant over the often unsuspecting farmer. Morris, with large investments in land, did not want change (Linklater, 2003).

Jefferson probably decided that discretion was the best part of valor in this case. Early in 1784 he knew that he would be going to France later in the year. Why risk his proposal for a decimal system of coinage by putting forward a grand scheme for weights and measures which incorporated his coinage proposal? No, it would be better to get a decimal system of coinage in place, and then, probably after his return from France, try for the grander scheme of weights and measures.

Jefferson lived in France between 1784 and 1789, and it is interesting that in 1793 the Revolutionary Convention established the Committee of Public Instruction and called on it to re-order education throughout France and, in particular, to rid the nation of all reactionary tendencies in education that had existed before the Revolution (Alder, 2002; Barnard, 2008; Farrington, 1910; Vignery, 1966). In 1795, France agreed that the decimal-based franc should replace the non-decimalized *livre tournois* that had existed before the Revolution. With respect to currency reform, the arguments used in Revolutionary France were similar to those used by Jefferson in the United States of America 10 years earlier.

Some Final Comments on the Achievement of a National Decimal Currency

It was not until July 1785 that Congress formally resolved that "the money unit of the United States of America be one dollar" (Boyd, 1953, p. 159), and it was not until October 1786 that it was agreed that a national mint should be established. In 1791, after Jefferson's return from Paris, Congress moved to activate their resolutions on the matter (Boyd, 1953; Linklater, 2003). On April 2, 1792, Congress passed the Coinage Act, which created the Mint and authorized construction of a Mint building in the nation's capital, Philadelphia (Goodwin, 2003). This was the first federal building erected under the Constitution.

The sequence of events which led to the adoption of a decimal system of currency in the United States of America has been well covered in history texts. But, what has gone largely unnoticed is how these events connected to school mathematics, for Thomas Jefferson was seeking to put in place an arithmetic for the people, one which ordinary people would be able to use confidently and competently when making written and mental calculations which involved money. He could not have made clearer his commitment to the need to work toward an arithmetic that all people could learn and use.

Throughout the sequence of events which resulted in Jefferson's, and not the Morrises', system of decimal currency being adopted, Jefferson's intellectual brilliance and political cunning were obvious. Although he would be the founder of a Republican Party that sought to preserve the powers of local communities and states on matters such as education, Jefferson recognized that it was important for the nation to make certain kinds of legislation in which national, rather than local or state, control was necessary. The introduction of a Federal decimal form of currency was one such matter, as was the related introduction of a national decimal system of weights and measures. Jefferson succeeded with his decimal currency proposal, but failed to garner sufficient support for his grand scheme for a decimal system of weights and measures. That failure will be looked at more carefully later in this book.

The identification of five features of Jefferson's big picture for currency reform, as they have been spelled out in this present chapter, make clear that Jefferson realized that school mathematics had an important role to play if his proposal for currency reform were to be adopted. Other writers (e.g., Conant, 1962) have drawn attention to Jefferson's proposals for education reform in Virginia, but ultimately his proposals for a three-level system of schooling in Virginia were never passed and would have little if any effect on schools—in Virginia, or elsewhere in the nation. But Jefferson's role in helping to get a system of decimal currency in place meant that his influence on the nation's education, and especially on school mathematics curricula, would be profound.

Interesting historical questions can be asked with respect to the introduction of the decimalized form of Federal money. Was Jefferson correct when he argued that most people would find it easy to carry out decimal calculations? How long did it take before all U.S. school children could cope with the decimal arithmetic associated with the new currency? Did, in fact, the introduction of a decimal system of currency mean that decimal fractions came to be more widely studied than they had previously been studied in school arithmetic? Was Jefferson's dream of a careful relationship between the national currency and a national system of weights and measures ever realized? Those questions will be addressed in later chapters of this present work.

An intriguing reflection is that although, arguably, Jefferson got the *big picture* right, so far as his proposals for a national currency and a national system of weights and measures were concerned, in his arguments for the small pictures (representing aspects of his big picture) he revealed his ignorance with respect to what went on in the name of school arithmetic. That was hardly surprising, given that Jefferson had had such a privileged upbringing, and there was no good reason why he should have known intimately the arithmetic that ordinary students learned, or did not learn, in ordinary schools—how, for example, they went about learning skills such as division of sterling currency. Jefferson's lack of detailed knowledge with respect to school arithmetic has not been pointed out by previous historians—and it is interesting that no-one has ever challenged the validity of what he said about the small-picture aspects. Perhaps, though, that is fortunate, because, ultimately, it was Jefferson's big picture which was important—for the national well-being and for school arithmetic.

References

Alder, K. (2002). *The measure of all things: The seven-year odyssey and hidden error that transformed the world.* New York, NY: The Free Press.

Allen, L. (2009). *The encyclopedia of money* (2nd ed.). Santa Barbara, CA: ABC-CLIO.

Barnard, H. C. (2008). *Education and the French Revolution.* Cambridge, UK: Cambridge University Press.

Bartlett, J. R. (1933). *Letter of instructions to the captain and the supercargo of the brig "Agenoria," engaged in a trading voyage to Africa.* Philadelphia, PA: Howard Greene and Arnold Talbot.

Bowditch, N. (1797). *Journal of a voyage from Salem to Manila in the ship Astrea, E. Prince, Master, in the years 1796 and 1797* (Handwritten manuscript held in the Bowditch Collection). Boston, MA: Boston Public Library.

Boyd, J. P. (Ed.). (1953). *The papers of Thomas Jefferson, March 1784 February 1785* (Vol. 7). Princeton, NJ: Princeton University Press.

Brock, L. V. (1975). *The currency of the American colonies, 1700–1764: A study in colonial finance and imperial relations.* New York, NY: Arno Press.

Conant, J. B. (1962). *Thomas Jefferson and the development of American public education.* Berkeley, CA: University of California Press.

Ellerton, N. F., & Clements, M. A. (2014). *Abraham Lincoln's cyphering book and ten other extraordinary cyphering books.* New York, NY: Springer.

Farrington, F. (1910). *French secondary schools: An account of the origin, development and present organization of secondary education in France.* New York, NY: Longmans, Green and Co.

Fauvel, J. (1999, April 15). *Thomas Jefferson and mathematics.* Lecture given at the University of Virginia.

Fenzi, G. (1905). *The rubles of Peter the Great.* Moscow, Russia: Open Library.

Frost, J. (1846). *Lives of American merchants.* New York, NY: Saxon & Miles.

Goodwin, J. (2003). *Greenback: The almighty dollar and the invention of America.* New York, NY: Henry Holt and Company.

Greene, J. P., & Jellison. R. M. (1961). The Currency Act of 1764 in imperial-colonial relations, 1764–1776. *The William and Mary Quarterly, 18*(4), 485–518.

Hepburn, A. B. (1915). *A history of currency in the United States with a brief description of the currency systems of all commercial nations.* New York, NY: The Macmillan Company.

Jefferson, Thomas to Edward Carrington, May 27, 1788 (in H. A. Washington (Ed.), *The writings of Thomas Jefferson*, New York, NY: H. W. Derby, 1861).

Jefferson, T. (1785). *Notes on the establishment of a money unit and of a coinage for the United States.* Paris, France: Author. (The notes are reproduced in P. F. Ford (Ed.), *The works of Thomas Jefferson* (Vol. 4, pp. 297–313). New York, NY: G. P Putnam's Sons. This was published in 1904).

Kleeberg, J. M. (1992). The New Yorke in America token. In J. M. Kleeberg (Ed.), *Money of pre-Federal America* (pp. 15–57). Proceedings of the Coinage of the Americas Conference, held at the American Numismatic Society May 4, 1991, in New York.

Linklater, A. (2003). *Measuring America: How the United States was shaped by the greatest land sale in history.* New York, NY: Plume.

Lossing, B. J. (1851). *The pictorial field-book of the Revolution*. New York, NY: Harper & Brothers.

Maier, P. (2010). *Ratification: The people debate the constitution, 1787–1788*. New York, NY: Simon & Schuster.

Massey, J. E. (1976). Early money substitutes. In E. Newman & R. Doty (Eds.), *Studies on money in early America* (pp. 15–24). New York, NY: American Numismatic Society.

McCusker, J. J. (1992). *Money and exchange in Europe and America 1600–1775*. Chapel Hill, NC: University of North Carolina Press.

Newman, E. P. (1990). *The early paper money of America*. Fairfield, OH: Krause Publications.

Nishikawa, S. (1987). The economy of Chōshū on the eve of industrialization. *The Economic Studies Quarterly, 38*(4), 209–222.

Notes on the Early History of the Mint. (1867). *Historical Magazine, 1*.

Ogg, F. A. (1927). *Builders of the Republic*. New Haven, NJ: Yale University Press.

Rappleye, C. (2010). *Robert Morris: Financier of the American Revolution*. New York, NY: Simon & Schuster.

Reid, J. P. (1991). *Constitutional history of the American Revolution, III: The authority to legislate*. Madison, WN: University of Wisconsin Press.

Seaman, W. H. (1902, March). How Uncle Sam got a decimal coinage. *School Science*, 232–236.

Sosin, J. M. (1964). Imperial regulation of colonial paper money, 1764–1773. *Pennsylvania Magazine of History and Biography, 88*(2), 174–198.

Vignery, J. R. (1966). *The French Revolution and the schools*. Madison, WI: University of Wisconsin.

Chapter 4
Weights and Measures in Teacher-Implemented Arithmetic Curricula in Eighteenth-Century North American Schools

Abstract: The decision having been made by the U.S. Congress to establish a decimal system of currency as the official national currency of the United States of America, one might have expected a decimal system of weights and measures to follow quickly. But that was not to be. Between 1784 and 1789 Thomas Jefferson was in France and therefore his role as principal catalyst for decisions on decimalization was muted. The chapter begins by placing the weights and measures decision in the context of economic and political forces operating within the new nation. From a lag-time theoretical perspective, although there were numerous available arithmetic textbooks which dealt with decimal fractions, the prevailing economic and political pressures negated any educational and mathematical pressures for more widespread applications of decimals. Thus, a curious result occurred—the United States decimalized its currency, but not its weights and measures.

Keywords: Ciphering books; Cyphering books; Decimal currency; Decimal fractions; George Washington; History of school mathematics; Implemented curriculum; Intended curriculum; Metric system; Weights and measures; Thomas Jefferson

Decimal Fractions in Author-Intended Arithmetic Curricula in British Schools

Decimal Fractions in Eighteenth-Century School Arithmetic Textbooks

Toward the end of the sixteenth century François Viète (1579), of France, and Simon Stevin (1585), of the Netherlands, demonstrated the usefulness of the concepts of decimal fractions for simplifying calculations associated with carrying out lengthy multiplications or divisions. Not long after that, with the introduction of logarithmic concepts in the early seventeenth century, it became clear that decimal fractions could be very useful in navigation, surveying and other practical situations. Since the commercially-oriented *abbaco* arithmetic curriculum dealt with the four operations on whole numbers, one might have expected any lag time before decimal fractions and logarithms were introduced into schools, to be short.

But, some believed that the concept of a logarithm was conceptually too difficult to be understood by most schoolchildren (Ellerton & Clements, 2012). This possibility meant that the lag time before a mathematical concept which had been developed by pure and applied mathematicians found its way into school mathematics curricula might be greater than expected because, many believed, most school-aged school children did not have have the mental capacity, or the mathematical preparation, to be in a position to deal with the concepts (Ellerton & Clements, 2014). Furthermore, there might not have been a sufficient body of available persons to teach the concepts effectively—however desirable it might have been that schoolchildren learn to understand and apply the concepts. Thus, for example, although youths might be shown how to use logarithms to calculate distances and angles in practical

M. A. (Ken) Clements, & N. F. Ellerton, *Thomas Jefferson and his decimals 1775–1810: Neglected years in the history of U.S. school mathematics*, DOI 10.1007/978-3-319-02505-6_4,
© Springer International Publishing Switzerland 2015

situations, because they were not able to *understand why* the procedures worked, they would not be well placed to apply them in potentially relevant situations.

Decimal fractions in Edmund Wingate's (1630) arithmetic. In fact, reasonably early in the seventeenth century authors of arithmetic textbooks for schools began to incorporate decimal fractions and logarithms into their intended arithmetic curricula. After groundbreaking publications by the British mathematicians John Napier (1614, 1619) and Henry Briggs (1617), Edmund Wingate, an English mathematician who was temporarily based in Paris, emphasized the power of the combination of decimal fractions and common logarithms—that is to say, logarithms to the base 10—to assist practitioners, such as surveyors, navigators, and carpenters, to make the kinds of calculations that they were likely to need to make in their daily workplaces (Wingate, 1624). On returning to England, Wingate (1630) wrote a text, designed for use in schools, in which he advocated the application of decimal fractions and logarithms as a way of simplifying calculations.

Wingate's courage paid dividends in the sense that even today his *Arithmetique Made Easie* is still recognized as one of the trend setters among early British arithmetics. In the 1680s, however, in John Kersey's (1683) revised edition of Wingate's textbook, decimal fractions were not formally dealt with until Chapter 22. When he first wrote his book, Wingate clearly *intended* decimal fractions to become an integral part of school arithmetic for *all* learners, but Kersey effectively allowed teachers in British schools to continue to follow the old, non-decimal, traditional sequence for elementary *abbaco* arithmetic. This meant that traditional, whole-number, teacher-implemented curricula would continue to be emphasized in the schools.

Decimal fractions in Edward Cocker's (1677, 1685) arithmetics. Other British authors of arithmetics adopted a similar approach to that of Kersey. For example, Edward Cocker's first arithmetic, published by John Hawkins in 1677, after Cocker's death, basically avoided decimal fractions. But, in 1685 Hawkins published *Cocker's Decimal Arithmetick* which was intended to serve the needs of those who went beyond the first book. In his preface, Hawkins's attitude was that, in taking this approach of presenting elementary *abbaco* arithmetic in two forms—one which did not acknowledge the potential usefulness of decimal fractions, and another which did—he was following the lead of Kersey (Cocker, 1685, p. vi). Cocker's *Decimal Arithmetick,* proved to be much less popular than his traditional non-decimal *Arithmetic.*

Decimal fractions in William Leybourn's arithmetics. William Leybourn (1626–1716) has been variously described as a mathematician, a publisher, a textbook writer, and a surveyor. He wrote numerous arithmetic textbooks designed for use in schools, and there can be no doubt that some of these were used in schools in the British colonies in North America (Ellerton & Clements, 2012). From the point of view of decimal fractions, Leybourn's (1690) *Cursus Mathematicus,* which had over 900 pages and brought together in two volumes much of his early writings on arithmetic, resembled Kersey's approach to decimal fractions insofar as it delayed the treatment of decimals until after he had dealt with "natural or vulgar arithmetic" (p. 1). Only then, in Part II, did he introduce "Decimal Arithmetick" (p. 85). But in his introduction to this section Leybourn showed much insight into forces influencing the adoption, or more explicitly, the non-adoption, of decimal arithmetic in schools. Leybourn (1690) wrote, somewhat prophetically:

Who was the first inventor of this decimal arithmetick, or brought it first to light, is hard to determine; but that it has been much improved since the invention of logarithms, all knowing artists cannot but acknowledge, and that at this day it is arrived to the zenith of its perfection. The excellency of this kind of arithmetick, will best appear when it shall be applied to mathematical practices; as in composing of tables of interests and rebates of money, in all sorts of measures, whether superficial or solid; in which cases (and many others, as in the subsequent treatises will at large appear), decimal operations do afford so great advantage that many ages before have not produced a more useful invention. And indeed this excellent kind of arithmetick might be yet farther improved, if all our coins, weights and measures (which are now divided into so many heterogeneous denominations) were divided and subdivided decimally. For by this means all calculations concerning trade would be performed with much ease and pleasure; especially the operations of multiplication and division: but these things last hinted are not in the power of any private person to effect, but wholly in such as are law-makers. (p. 85)

This remarkable statement is clearly consistent with the lag-time theoretical perspective outlined in Chapter 1.

Leybourn (1690) argued that despite the good work of researchers and of applied mathematicians, the general adoption of decimal arithmetic into everyday practices, and presumably into schools, would not occur until decimalization had become government policy with respect to coinage and to weights and measures. Analysis of the teaching and learning of decimal fractions in British and North American schools in the seventeenth and eighteenth centuries provide strong support for Leybourn's contention.

The Decision Not to Decimalize U.S. Weights and Measures

Between 1790 and 1796 President George Washington, Thomas Jefferson and James Madison were among those who urged Congress, to move toward developing and adopting a uniform Federal decimalized system of weights and measures. Ultimately, though, the opportunity for a new system was lost. The momentous decision of Congress not to accept the decimalized system of weights and measures put forward by Jefferson, and favored by Washington and Madison, has been ably documented by Julian Boyd (1961) and Andro Linklater (2003), and details related to that decision will not be given here. But three aspects relevant to the research questions being addressed in this book need to be noted.

Lag-Time Considerations and a Window of Opportunity

The theoretical perspective provided by the lag-time concept and Figure 1.3, in Chapter 1 of this book, predicted that decimal fractions would not be seriously studied in schools unless and until social, economic and political circumstances were conducive to implemented arithmetic curricula in North American schools embracing decimal arithmetic as an essential core component of "arithmetic for all." Towards the end of the eighteenth century a form of curriculum inertia was in place—teachers, who controlled implemented curricula, would not consciously agree to ask students to learn decimal fractions unless they were persuaded that decimals were core knowledge, essential for all persons preparing cyphering books. That was unlikely to happen because local forms of sterling currency continued to be used, and the traditional units for weights and measures remained in place.

As Linklater (2003) has shown, in 1790 the U.S. debt was huge, the public purse having been emptied and an enormous national debt created during the Revolutionary War. The most obvious potential economic savior for the nation was a large-scale sale of land west of the Ohio River, including land west of the Mississippi River—and there were two important pre-conditions for such a sale. The first was that agreements needed to be reached with indigenous Americans who had occupied the land for centuries; and the second was that the vast land tracts could not be sold before they had been surveyed. The political challenge of reaching agreements, and the physical task of surveying needed to be done quickly. There was only a short window of opportunity.

For many members of Congress, and especially those with investments in land, it made little political or economic sense to introduce a system of weights and measures that rendered all existing surveying maps out of date, which would threaten to make many of the qualified and professional surveyors feel incompetent, and which promised to make existing surveying instruments obsolete. The arguments put forward by those who opposed decimalization of weights and measure were as simple as they were compelling—a Federal decision to decimalize weights and measures would slow down, dramatically, the flow of land money into national coffers. From an education perspective, it would de-skill teachers of arithmetic, many of whom would have been even more unprepared than usual to teach arithmetic because existing textbooks and cyphering books would not have been appropriate.

Thus, although the new decimal coinage system had been accepted by Congress—because the new nation had little choice but to introduce a currency system that differed obviously from British sterling—a radically new decimal system for weights and measures was regarded by many as dangerous from economic and education perspectives. It was easy for Congress to do what it did do—reject it, on the grounds that it would have been unwise to do otherwise at that time. But Washington, Madison, and Jefferson, and some others, wanted to go ahead with the decimalization approach, despite the arguments against such a course of action.

Washington, Madison, and Jefferson seemed to recognize that it might be "now or never." They sensed that there would be something incongruous about decimalizing currency but not weights and measures. But, democracy would prevail (Jefferson, 1785).

In 1790, Jefferson put forward, as an alternative to his radical proposal to decimalize weights and measures, a non-decimalized system of weights and measure which, although it would have standardized length and time by reference to the second's pendulum, was far less radical than his decimalized scheme. Jefferson seemed to sense something mystical, something from very high antiquity, about his decimal scheme, and he recognized that the new dawn of the period 1775–1790 provided a golden opportunity for its introduction.

Although, Robert Morris, the Financier, whose thinking was always in line with the needs of land speculators, was wrong—and he probably knew he was wrong—when he claimed that weights and measures were standard throughout the Union, his 1782 statement that that was indeed the case probably represented well the feelings of those with money and large investments in business enterprises. From their perspective, it would be counterproductive to change a system of weights and measures that was allowing speculators and merchants to prosper.

Jefferson's Leadership on Weights and Measures: An International Perspective

Thomas Jefferson's system of weights and measures had an elegant yet deeply logical structure. When considering the proposals for establishing a decimalized system of weights and measures, many members of Congress would not have been fully aware of the opportunity with which they, and the young nation, were being presented. Basic units were to be defined, one after the other. First, would come the unit for time—a second, would be defined in relation to the rotation of the earth, and that definition would be incorporated into the "seconds-pendulum" definition of a unit for length (the American foot). Then, the unit for length would become logically part of the definitions for area, and volume; then the definition of capacity would be defined in terms of volume, and assumed in the definition of weight, which would be used in the definition of monetary value (inherent in Jefferson's system of decimalized coinage).

Jefferson's "Some Thoughts on a Coinage" (circa March 1784, in Boyd, 1953, pp. 173–175) made clear that *before* he went to France in 1784, the principal writer of the Declaration of Independence had developed a radical and coordinated system of weights and measures which linked time, length, area, volume, capacity, weight, and coinage. Then, while Jefferson was in France, his scheme was refined as a result of numerous discussions that he had with the Frenchmen Charles Maurice de Talleyrand-Périgord, (usually known as Talleyrand) and Jean-Antoine Nicolas de Caritat, Marquis de Condorcet (usually known as Condorcet), with Sir John Riggs-Miller, from England, and with Benjamin Franklin (Boyd, 1961). Although the origins of the metric system are traceable to a system of weights and measures proposed in England in 1668 by Bishop John Wilkins (1668), many of the main features of the "metric system" of weights and measures introduced by the French in the 1790s had been the subject of discussion among savants for over a century, and one should not underestimate the polish that might have been added to the metric system as a result of Talleyrand's and Condorcet's interactions and conversations with Jefferson.

The history of the decimalization of weights and measures certainly did *not* start with Gabriel Mouton (1670), the French scientist, or with the Dutch mathematician and astronomer, Christiaan Huygens (see Thomson, 1874). Nor did it end with the development and adoption of the metric system by French mathematicians during the French Revolution. A letter by an anonymous author which appeared in *The Philosophical Magazine* ("Wright on measuring the meridian—Wright, Wren and Wilkins on an universal measure," 1805) noted that the French scientists and mathematicians seemed to be unwilling to acknowledge Wilkins's (1668) publication. The letter also noted that the French wanted to give Mouton and Huygens credit at the expense both of Edward Wright who, in 1599 had proposed using the earth's meridian as a standard, and Wilkins, who had proposed a measurement system based on the seconds pendulum.

The system of weights and measures introduced in France during the 1790s differed in its fundamental structure from Jefferson's preferred system. Jefferson wanted the unit of length to be defined with respect to a pendulum-like rod but the French Revolutionaries defined their unit of length as part of a meridian of longitude which passed through Paris (Alder, 2002).

Decimal Fractions, Author-Intended and Teacher-Implemented Arithmetic Curricula

Most eighteenth-century British authors of school arithmetics (e.g., Cocker, 1677; Kersey, 1683; Leybourn, 1690) included chapters on decimals, with some authors pointing out how a decimalized form of numeration could not only revolutionize school arithmetic but also have major applications. However, with only a few exceptions (e.g., Wingate, 1630), the texts did not present decimal arithmetic until after the total sequence of elementary *abbaco* arithmetic (from numeration and the four operations, to the rules of three, to progressions) had been dealt with using traditional whole-number approaches. If an author went beyond that tradition, there would usually be a section on "vulgar arithmetic" before the section on decimal arithmetic. That was the case, for example, with the two most popular arithmetics—those by Francis Walkingame and Thomas Dilworth—used in British schools during the period 1750–1850.

British historians (e.g., Denniss, 2012; Michael, 1993; Stedall, 2012) have pointed to the unparalleled popularity, with respect to school mathematics in Great Britain between 1750 and 1850, of Francis Walkingame's *The Tutor's Assistant; being a Compendium of Arithmetic and a Complete Question Book.* A 1785 edition of Walkingame's book devoted its first 97 pages to "Part I: Arithmetic in Whole Numbers," then the next 10 pages to "Part II: Vulgar Fractions," and then the next 42 pages to "Part III: Decimals." The second most popular text was that by Thomas Dilworth's *Schoolmasters Assistant*, a text which was widely used in North America in the second half of the eighteenth century. Dilworth's book, which was dedicated "to the revered and worthy schoolmasters in Great Britain and Ireland" (Dilworth, 1762, p. iii), devoted its first 110 pages to "Part I: Of Whole Numbers," then the next 12 pages to "Part II: Of Vulgar Arithmetic," and finally the next 46 pages to "Part III: Of Decimal Fractions."

The textbooks suggested that decimal arithmetic should be part of the intended curriculum for teachers and students. But, did Walkingame and Dilworth, and numerous other writers whose arithmetics presented whole number arithmetic, vulgar fraction arithmetic, and decimal fraction arithmetic as separate areas of study, include the sections on decimal fractions merely to catch the small number of students who might continue to study arithmetic beyond whole numbers and vulgar fractions? What proportion of those teachers who required their students to use texts like those by Walkingame and Dilworth ever got around to asking their students to study the chapter(s) on vulgar arithmetic or decimal arithmetic? Such questions relate to the "implemented curriculum," and answers will be given later in this chapter, and in the next two chapters, with respect to situations in Great Britain and North America.

With two exceptions (e.g., Greenwood, 1729; Grew, 1758), between 1607 and 1776 the only mathematics textbooks available for use in the British colonies in North America were those of European origin. In fact, most of the school arithmetics used in North American schools during that period had British authors. Among the most popular arithmetics used in North American schools were those written by Edmund Wingate, Edward Cocker, George Fisher, Thomas Dilworth, James Hodder, and Francis Walkingame (Ellerton & Clements, 2012). Some of these were reprinted in North America, but in such cases the North American editions were merely copies of British editions. Furthermore, the teachers who taught the North American students were often from Great Britain. Thus, the intended curricula for school arithmetic in the two nations would be expected to have been similar, and one might therefore expect the implemented arithmetic curricula in the two nations to have been similar.

However, the very different conditions within the various American colonies could have resulted in different implemented curricula. Thus, there is a need to study North American cyphering-book data for the period before 1792, especially with respect to decimal fractions. To what extent did the cyphering books include entries on the arithmetic of decimal fractions, including possible applications in applied contexts such as arose with tasks in the content areas known as compound operations, reduction, rules of three, exchange, practice, loss and gain, discount, tare and tret, and simple and compound interest? In order to investigate that question, those cyphering books in the Ellerton-Clements collection which were prepared before 1792 were studied.

Weights and Measures Tasks in Implemented (Cyphering-Book) Arithmetic Curricula in Pre-Revolutionary North American Schools

Money was certainly not the only aspect of applied arithmetic emphasized in eighteenth-century school arithmetic, and in this section entries in North American cyphering books related to weights and measures will be the focus of attention. All elementary arithmetic textbooks, and most cyphering books, paid attention to how quantities involving length, area, volume, capacity, weight, time, and angle, could be added, subtracted, multiplied and divided. Both textbooks and cyphering books also had extensive sections on "reduction" and the "rule of three" in which calculations with respect to weights and measures could also be involved.

A Note on the Mathematical Basis of Compound Operations Tasks

Before reproducing pages from cyphering books showing calculations with respect to weights and measures, it will be in order here to comment on the mathematical basis of the calculations. Suppose three towns, A, B, and C are on a straight road, with B being 10 miles west of A, and C being 15 miles west of B. Then one could conclude that C is $(10 + 15)$ miles, or 25 miles, west of A. Colloquially, one might talk of "adding" 15 miles to 10 miles, to get 25 miles. From a mathematical perspective, however, one adds *numbers* but *not distances*. Similarly, if one were calculating the area of a rectangular field 200 yards long and 300 yards wide, one might say that the area is (200×300) square yards. However, the lengths of two adjacent sides of the rectangle were not multiplied—rather, the length *measures* of (i.e., the numbers associated with) two adjacent side lengths were multiplied, and the appropriate unit for area then attached to the calculation.

In the *abbaco* tradition the processes associated with "adding," "subtracting," "multiplying," and "dividing" quantities whose extents were expressed in traditional units was known as "compound operations." Students typically showed how to carry out "compound additions," etc., in their cyphering books. Although one can easily become familiar with the algorithms required to complete the addition, or subtraction, or multiplication or division of the amounts, from a purely mathematical perspective, $+$, $-$, $\times$ and $\div$ are operations defined on, and applied to, real numbers. So, in essence, the idea of adding and subtracting amounts of quantities requires an extension to the conceptual base. If one can add or subtract quantities, then one would expect to be able to multiply and divide them too. And, indeed, there were algorithms available to carry out each of the "compound operations." Thus, for example, in a topic traditionally called "cross multiplication" (and, sometimes called "duodecimals"), amounts of quantities were multiplied and divided, and sensible answers relating to real-life situations obtained (see, e.g., Pike, 1788).

Cornelius Houghtaling's (1775) Entries on Weights and Measures

In this section various entries made by Cornelius Houghtaling in his cyphering book on topics related to weights and measures will be reproduced, and then comments will be made on the educational significance of those entries. The extent to which Cornelius's entries on weights and measures were similar to, or different from, entries on weights and measures in other North American cyphering books of the period will also be considered.

Table 2.2, in Chapter 2, revealed that Cornelius Houghtaling included entries in his cyphering book on both vulgar and decimal fractions, as well as on weights and measures. Cornelius was an 18-year-old youth living in and around New Paltz, a Huguenot settlement located north of what is now the city of New York. In 1775, when Cornelius began to prepare his cyphering book, he was a volunteer in the Revolutionary Army, and skirmishes between British soldiers and volunteers were taking place in and around New Paltz.

Cornelius Houghtaling (1775) on "compound operations." Cornelius included examples concerned with compound addition, subtraction, multiplication and division for money, troy weight, apothecaries weight, avoirdupois weight, wine measure, beer and ale measure, dry measure, long measure, land measure, cloth measure, and time. Compound operation tasks on the three different types of weight categories—troy, apothecaries, and avoirdupois—were considered separately, and different units, and unit relationships which needed to be known, were summarized. Cornelius's book also referred to how measures of capacities (that is to say, amounts of fluid that a container can hold) depended on the fluid under consideration—thus, for example, there were different units and relationships for wine measure and for beer and ale measure.

Cornelius and other youths in New Paltz would have had experience handling goods that came to the town via the Wallkill River, and Figure 4.1 (taken from a *Wikipedia* article on "English Wine Cask Units"—see http://en.wikipedia.org/wiki/English wine_cask_units) shows measures and units associated with barrels commonly used in North America around 1775. In 1707 the British parliament had defined a *wine gallon* to be 231 cubic inches—which, although it differed from the imperial gallon, had been adopted by the United States as its customary gallon. A *rundlet* was, by definition, the same as 18 wine gallons or one-seventh of a *wine pipe*, and a *wine barrel* as $31\frac{1}{2}$ wine gallons or half a *wine hogshead*. A *tierce* was 42 wine gallons, which was half a *puncheon* or one-third of a *wine pipe*. A *wine hogshead* was two wine barrels or 63 wine gallons, or one-fourth of a *wine tun*, and two wine hogsheads, three tierce, or seven rundlets made up one wine pipe. A wine *tun* was two wine pipes, three puncheons, or 252 wine gallons. All of these various terms relating to capacity often appeared in North American cyphering books.

Figure 4.1. Common units used in eighteenth-century Great Britain and North America (from http://en.wikipedia.org/wiki/English_wine_cask_units).

Figure 4.2 reproduces a page from Cornelius's cyphering book showing different units (Grains, Scruples, Drams, Ounces and Pounds) for "apothecaries weight." Note that pharmacists ("apothecaries") compounded their drugs using apothecaries weight units, but sold them in avoirdupois units. Figure 4.3 reproduces the first of two pages showing "Reduction Tables." The statements under the heading: "Avoirdupoise Wt." "Tuns mult by 20 is Hundreds"; "Hundreds mult by 4 is Quarters"; "Quarters mult by 28 is Pounds"; "Pounds mult by 16 is Ounces"; and "Ounces mult by 16 is Drams" could be misleading. The truth or falsity of those statements depended on how they were interpreted. Presumably, the intention was to mean, for example, "The *number of* tuns multiplied by 20 gives the *number* of hundreds," but for someone who did have any idea of the meanings of "tuns" and "hundreds" in this context, the statement "Tuns mult by 20 is Hundreds" might have suggested that Hundreds were 20 times as big as Tuns, which was not the case. All of the statements shown in Figure 4.2 suffered from this same ambiguity.

Figure 4.4 shows how Cornelius "added" quantities of wine." From a purely mathematical perspective, the apparently simple compound addition tasks shown in Figure 4.4 involved changing bases when carrying out additions. Thus, for example, with the exercise on the left side, 5 pints needed to be converted to 2 quarts and 1 pint; then, 14 quarts needed to be converted to 3 gallons and 2 quarts; and 243 gallons to 3 hogsheads and 54 gallons; and 15 hogsheads to 3 tuns and 3 hogsheads. A check was shown, whereby the sum of the second, third, fourth and fifth amounts was added to the first amount, to get the total amount. Details of another quantity measure, this one associated with so-called "dry measure," could also be found in Cornelius' book. Measurement of length was called "long measure," but the term "area" was not used—instead, there were "land measures" and "cloth measures." The term "volume" was not used, but "solid measure" was.

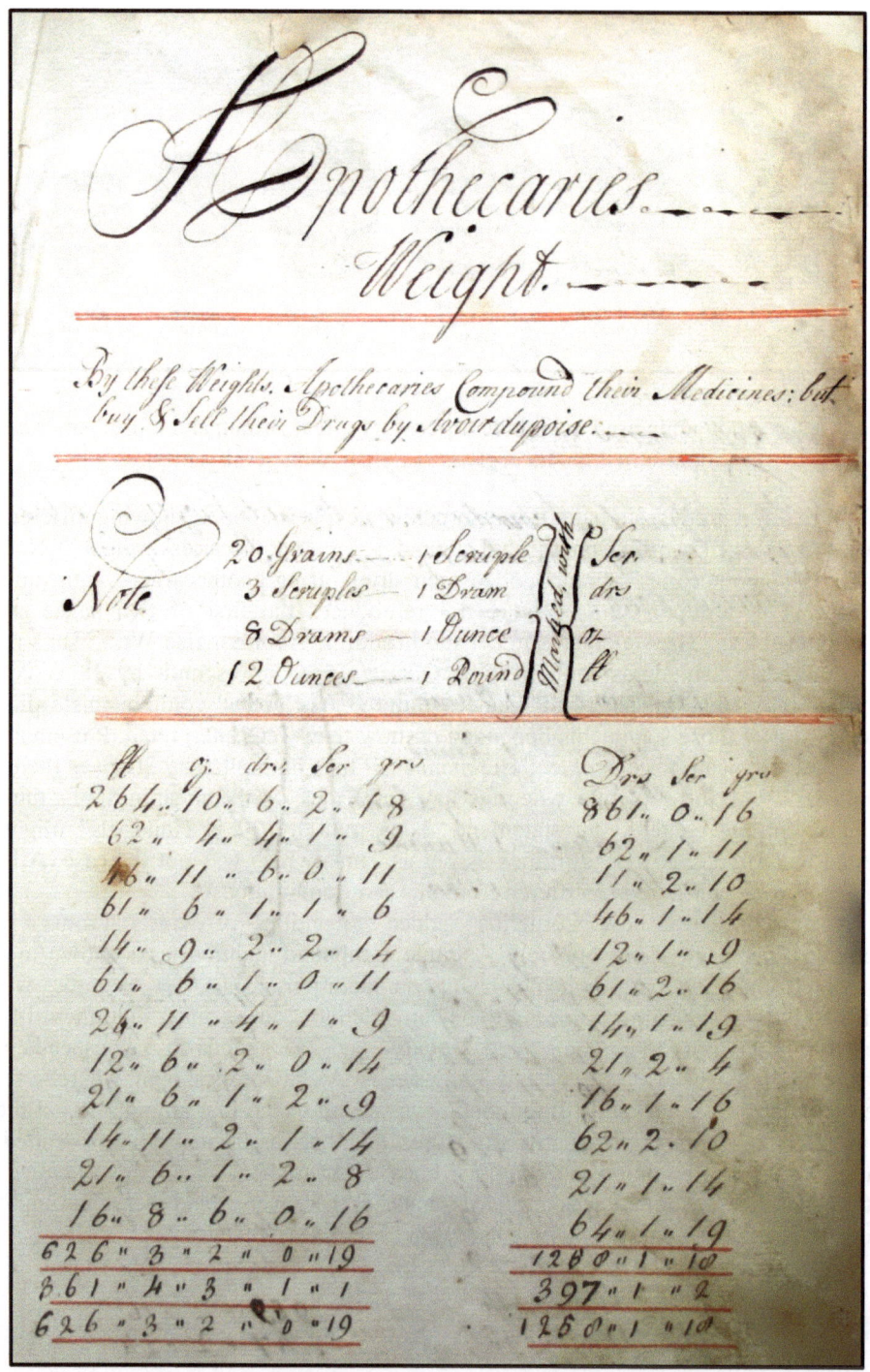

Figure 4.2. Units for apothecaries weight (Cornelius Houghtaling, 1775).

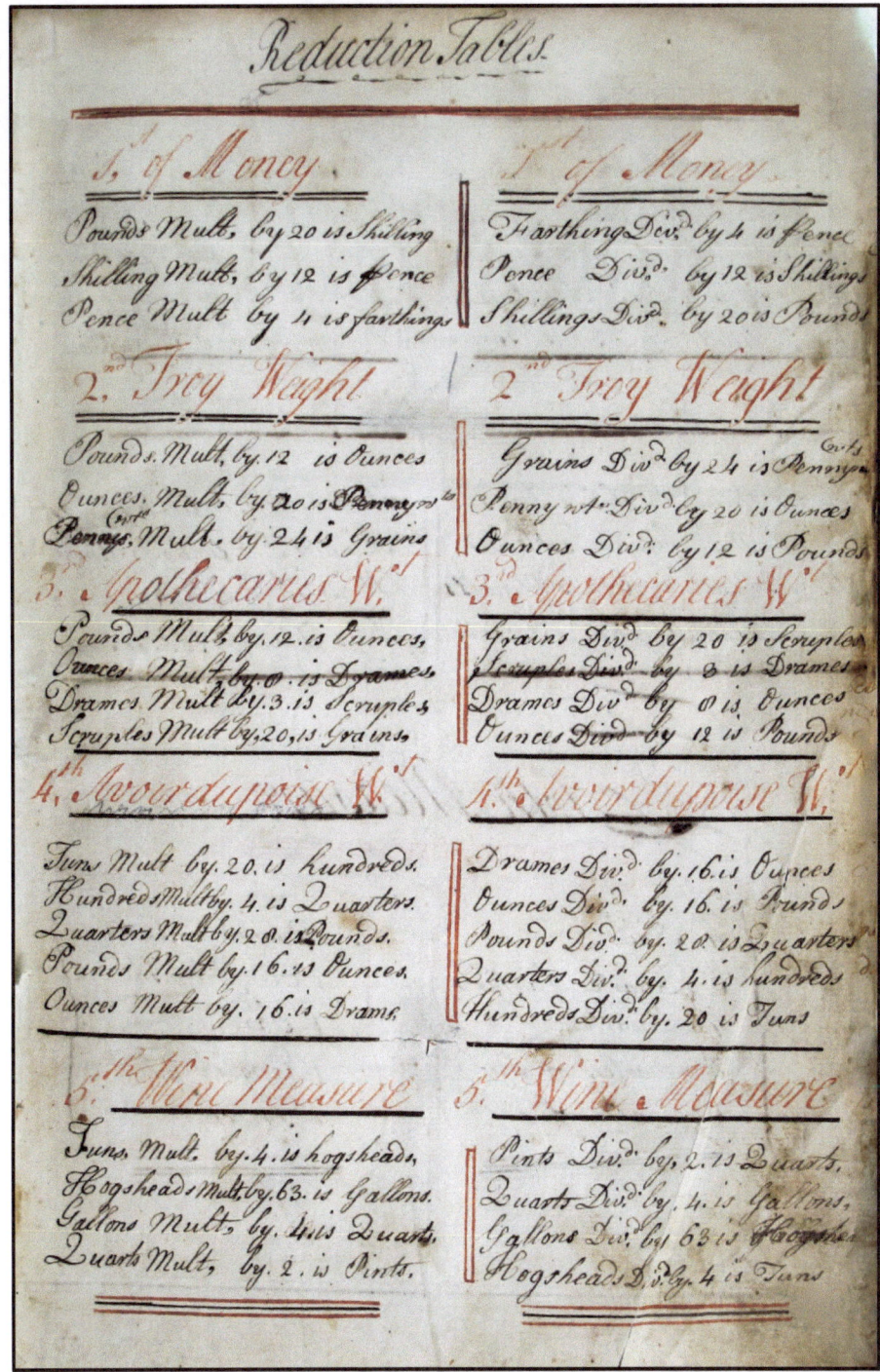

Figure 4.3. "Reduction" tables in Cornelius Houghtaling's (1775) cyphering book.

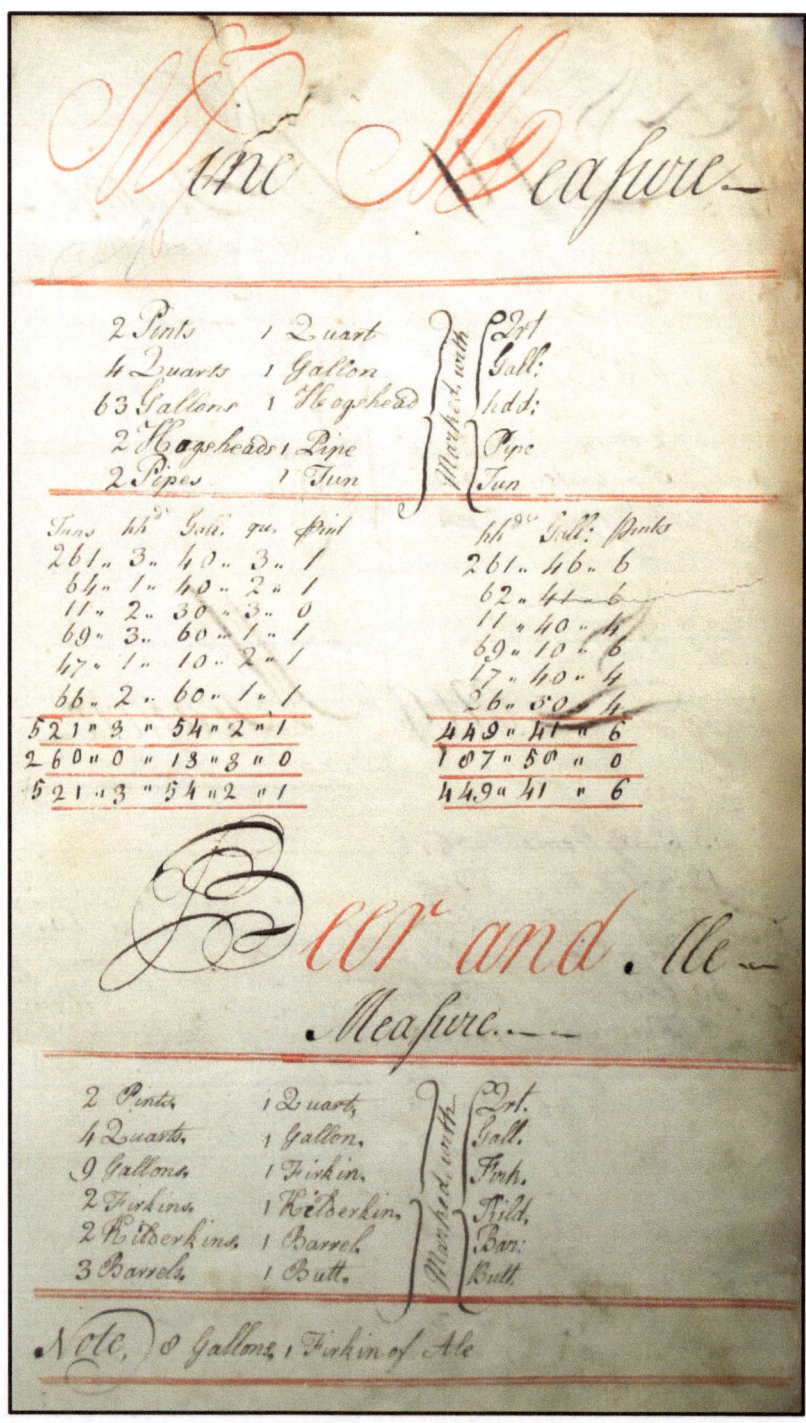

Figure 4.4. Cornelius Houghtaling (1775) "adds" wine, and beer and ale measures.

Many units could have been, but were not, named in the two pages of "Reduction Tables" in Cornelius's cyphering book (see Figure 4.3). Of the first 36 pages in Cornelius's cyphering book, 33 were dedicated to simple tasks involving the combinations of different quantities. This part of *abbaco* arithmetic was commonly termed "compound operations," and those preparing a cyphering book for the first time—usually, children of about 10 years of age—would deal with "compound operations" after they had dealt with the topics "numeration," "addition," "subtraction," "multiplication," and "division." Sometimes (as was the case with Cornelius), compound-operation tasks associated with a particular operation would be dealt with immediately after "pure-number" exercises on that operation had been completed, and before the next "pure-number" operation was introduced. In the first 10 pages of Cornelius's cyphering book, altogether 63 different units for these quantities were mentioned.

With respect to Cornelius's working for the two exercises shown in Figure 4.3, two additional observations should be made:

1. Cornelius showed checks for his working. The first "answer" for each exercise showed the sum requested; that was followed by the sum of all the amounts except for the first line; and then, as a check, this last amount was added to the first-listed amount.

2. With the first exercise, "pipes" were not mentioned, so one needed to be aware that there were four hogsheads in a tun.

It should not be assumed that the kind of addition tasks that Cornelius Houghtaling completed in Figure 4.4 were so easy that all students would have had no difficulty completing them and understanding the mathematics behind them. In eighteenth-century North America there was a gap between the intuitive forms of hornbook education offered to young children in dame schools and the knowledge of reading, writing and arithmetic that was needed by 10-year-olds who would begin to create their own cyphering books. Many young children who began to prepare an elementary cyphering book had little understanding of what they were doing (Earle, 1899). The Reverend Warren Burton (1833) recollected that in 1812, at the age of 12, he had been perplexed when he first began to prepare a cyphering book at a small, one-room school house in New England:

> I met with no difficulty at first; simple addition was as easy as counting my fingers. But there was one thing I could not understand—that carrying of tens. It was absolutely necessary, I perceived, in order to get the right answer; yet it was a mystery which that arithmetical oracle, our schoolmaster, did not see fit to explain. It is possible that it was a mystery to him. Then came subtraction. The borrowing of ten was another unaccountable operation. The reason seemed to me then at the very bottom of the well of science; and there it remained for that winter, for no friendly bucket brought it up to my reach. … Each rule was, or rather was to be, committed to memory, word for word, which to me was the most tedious and difficult job of the whole. (pp. 115–116)

Burton (1833) added that it took him three winters before he reached the rule of three.

Cornelius Houghtaling (1775) on "reduction ascending and descending." After completing a section on compound operations tasks which were concerned with the bewildering array of units and relationships associated with money, weight, length, area, volume, and time, Cornelius moved on to the topic "reduction." He began with a definition of "reduction ascending and descending" (see Figure 4.5):

> Reduction ascending and descending teacheth to bring numbers of one denomination into another by multiplication or division as the case requireth. All great names are brought into small by multiplication and all small names into great, by division. To change one sort of money, weight, measure, &c., into another, reduce both into one name and divide the one by the other.

The wording of this "definition" of reduction was checked against many popular arithmetics available in North America in the 1770s and it was not found in any of them. It would appear likely that the words were dictated to Cornelius by a tutor, or were copied from an older cyphering book.

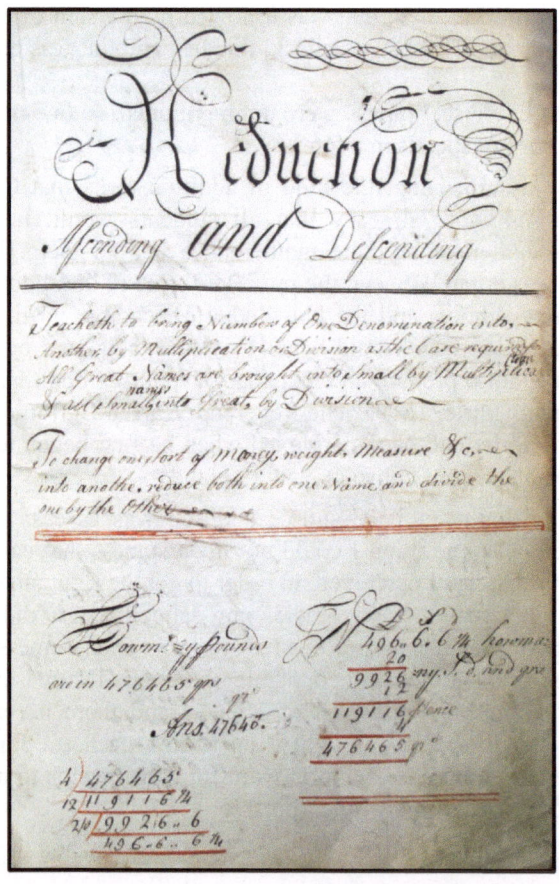

Figure 4.5. Cornelius Houghtaling's (1775) definitions for reduction.

Following this definition, Cornelius quickly moved on to the task "How many pounds are in 476465 qrs." Although the question did not make it obvious, this was a question requiring 476465 farthings (denoted by "qrs") to be converted to pounds, shillings, pence and farthings. Cornelius quickly reached the answer by dividing by 4, by 12, by 10 and by 2— with the remainders requiring interpretation in order to arrive at the correct answer. After the division by 4, 119116 and 1 remainder was obtained, which was expressed as 119116¼; then, after division by 12, 9926 and a remainder 6 was obtained (the remainder should have been 4). Cornelius wrote 9926,, 6. After division by 20, the incorrect 496,,6,,6¼ was obtained. Cornelius then proceeded to convert 496 pounds, 6 shillings, 6 pence and 1 farthing back to farthings, and incorrectly obtained the original 476465 farthings. So, he had demonstrated the two types of reduction, with one set of calculations serving as a check.

Cornelius then tackled other reduction tasks which involved different quantities. One question asked how old someone born in the year 1757 would be in minutes, given that there were 13 months 1 day and 6 hours in a year. In 1775, it was usual to divide a year into 13 months 1 day and 6 hours—the idea was said to have originated from Julius Caesar, and the term "Julian calendar" was common. Cyphering books included many such pseudo-reality questions—for example, one was likely to find questions in cyphering books asking how many minutes had passed from the creation of the earth, which was given as having occurred 4004 years before the birth of Jesus Christ. A slightly more "real" question was expressed in the following whimsical way:

Cornelius. I desire to know how many times a wagon wheel which is 18 feet 6 inches round will turn from here to your house, supposing it to be 2 miles.

After converting 18 feet 6 inches to 222 inches, Cornelius then multiplied 2 (miles) by 8 to get 16 (furlongs), which, when multiplied by 220, then 3 then 12, gave 126720 (inches), which when divided by 222 yielded an answer of approximately "570 turns round."

Figure 4.6 shows Cornelius's setting out for four reduction tasks: "In 8700 English miles, how many furlongs, yards, feet, inches and barley corns?" "How many English miles are in 1653696000 barley corns?" "How many furlongs are in 3726030 inches?" And, "In 470 furlongs, 100 yards, 2 feet 6 inches, how many yards, feet and inches?" All of Cornelius's calculations were beautifully set out, using two ink colors and high-quality calligraphy. Although there were no direct explanations of why calculations were performed, anyone who knew and understood the relationships involved would have been able to follow the working.

Cornelius Houghtaling (1775) and proportional thinking. After the section in Cornelius's book on reduction came a lengthy section on proportional thinking, for which the various rules of three needed to be applied. Although there were many questions which involved sterling currency calculations, there were also many questions which were mainly concerned with weights and measures. Some of these questions were:

1. If 12 men build a wall 30 feet long and 6 feet high and 3 feet thick in 15 days, in how many days will 60 men make a wall 300 feet long, 8 feet high and 6 feet thick?
2. If 3 men can reap 96 acres of wheat in 36 days, I demand how many acres 64 men may reap in 48 days.
3. How many yards of matting that's ½ a yard wide is enough to cover a flour (*sic.*) that's 16 foot wide and 20 foot long?

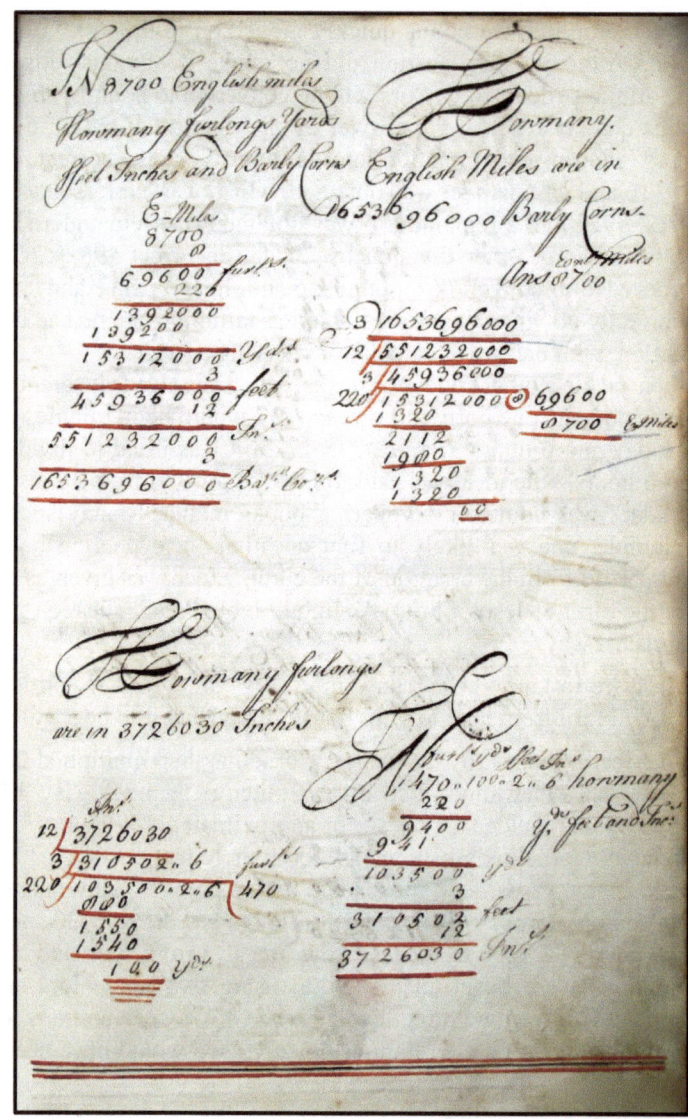

Figure 4.6. Cornelius Houghtaling's (1775) calculations for four reduction tasks.

Cornelius's working for two inverse rule-of-three tasks are shown in Figure 4.7. The tasks were:

1. Captain is besieged with 160000 men which he has only provision for 237 days. How many of these men must he disband to make the said provision last 2 years and ½?

2. A regiment of soldiers consisting of 1000 are (*sic.*) to have new coats and each containing 2 yards 2 quarters of cloth that is 5 quarters wide and they are to be lined with shalloon that's 3 quarters wide. I demand how many yards of shalloon will line them?

Cornelius's responses to the inverse rule-of-three tasks shown in Figure 4.7 featured what Ellerton and Clements (2012) called the "problem-calculation-answer" (PCA) genre—many calculations were shown but virtually no explanations were given for why the calculations were made; then, the answer was stated. In the "Captain besieged" problem, 237 was multiplied by 2 to get 474, then 160000 was multiplied by 474 to get 754840000; then $2\frac{1}{2}$ was multiplied by 365 to get $912\frac{1}{2}$, which was multiplied by 2 to get 1825. Then, 54840000 was divided by 1825, giving 41556 which, when subtracted from 160000 gave 118444. This was specifically stated as the answer to the question (immediately after the statement of the question).

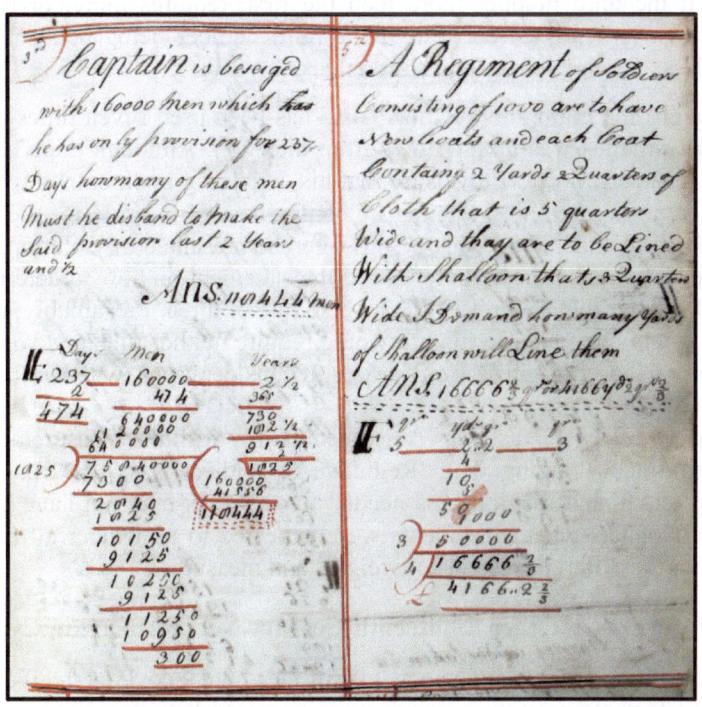

Figure 4.7. Cornelius Houghtaling's (1775) solutions to inverse rule-of-three problems.

The calculations for the "Captain besieged" problem were neatly and accurately performed. But no explanation was given for why the 237 was multiplied by 2. If one looks carefully, one can find that "2 years and a ½" was converted to 1825 half-days, so the unit of time being assumed in the solution would appear to have been a "half-day," which explains why the 237 was multiplied by 2. After the long division of 75840000 by 1825 a remainder of 300 was obtained, but no explanation was offered for why that remainder was deemed to be irrelevant to the solution.

The solution given to the "Captain besieged" problem raised the question "How did Cornelius know to do what he did?" The answer to that question could lie in the fact that on the previous page in the cyphering book Cornelius had not only specifically written down the rule for solving "inverse rule of three" tasks, but had also stated how a problem-solver might determine whether a given proportion task was an "inverse-rule-of-three" task rather than a "direct-rule-of-three" task. On July 29, 1775, Cornelius had written, under an extravagant calligraphic heading "The Rule of Three Inverse," in an equally extravagantly form of penmanship:

> In this rule questions are stated and the numbers prepared. In the same manner as in the Direct Rule: but the (*sic.*) differ In (*sic.*) that hear (*sic.*) you must multiply your first and second numbers together and divide the product by the third the (*sic.*) quotient as before is the answer.

> To know whether a question belongs to the direct, or inverse rule of three: Observe if the third number more than the first, requires more; or being less, requires less: tis (*sic.*) Direct. But if the third number being being (*sic.*) more requires less, or being less requires more, tis (*sic.*) inverse.

In these paragraphs the Latin abbreviation "*sic.*" has been used seven times—that was done to suggest that Cornelius was taking notes from dictation. Although there can be no doubt that Cornelius made a real effort to ensure that his cyphering-book entries were as attractive as possible, the question remains whether he comprehended the rule for the inverse rule of three, and whether he understood what he did when he presented his solution. An even more important issue was whether he comprehended the statement on how to determine whether a task required the direct rule of three or the inverse rule of three. Presumably, many situations that Cornelius would meet later in his life would require neither rule, and whether Cornelius would have been able to recognize, from structural considerations, whether that was the case, can never be known.

The "Captain besieged" task was concerned mainly with the measurement of time, and associated proportional thinking. The "Regiment of soldiers" task was concerned with the measurement of cloth, and relationships needed for that task may not have been familiar to Cornelius. One wonders whether students were expected to remember all the relationships between units for the different systems of weights and measures.

Weights and Measures in Other Eighteenth-Century North American Cyphering Books

Table 2.2 in Chapter 2 offered an analysis of the content covered in 20 cyphering books in the Ellerton-Clements collection which were prepared between 1742 and 1792. It was found that about three-fourth of the pages in the 20 handwritten manuscripts were dedicated to either money or to weights and measures, with over 30% of the pages mainly concerned with non-money tasks involving weights and measures. These tasks mostly required the application of compound operations, or the methods used for reduction, or the various rules of three.

In some eighteenth-century cyphering books almost all of the pages were concerned with either money tasks or weights and measures tasks. Thus, for example, between 17 and 18 of the 22 pages of Jonathan Livermore's (1742–1743) cyphering book were concerned with the addition, subtraction, multiplication and division of cloth measures, troy weights, avoirdupois weights, dry measures, wine measures, long measures, and time, or with tasks involving reduction of money and of weights and measures, or the application of the direct rule or the rule of three to money or to weights and measures tasks. Lists of unit relationships were given (e.g., for long measure, "for 3 barley corns 'carry' 1 inch, for 12 inches carry

1 foot, for 16 feet and ½, carry 1 rod, for 40 rods carry 1 furlong, for 8 furlongs carry 1 mile, and for 60 miles carry 1 degree"). Most of the compound operations tasks were done vertically, using "carrying." Sources of common misconceptions appeared in statements like "pounds multiplied by 20 are shillings." Reduction problems appeared in the form, "In 360 degrees how many miles, furlongs, rods, feet, inches, and barley corns?" After one set of PCA calculations it was concluded that there were "4105728000 barley corns round the earth."

It should not be assumed that the apparently strong emphasis on purely practical arithmetic was only to be found in cyphering books prepared by students from struggling families. The record indicates, for example, that the Jonathan Livermore who prepared the 1742–1743 cyphering book was born into a well-to-do family at Northboro, Massachusetts in December 1729, that he graduated from Harvard College in 1760, and that for 15 years (1763–1777) he was a Congregational pastor, after which he became a saw-mill owner (Cutter, 1914; Stafford, 1941).

The Challenge to Standardize, and Decimalize, Weights and Measures

Much has been written about the large number of units employed for weights and measures in the North American colonies during the eighteenth century. The most commonly-used units had British origins—but, German-background persons (many of whom were in Pennsylvania), and Dutch-background persons (many of whom were in and around New York) often preferred to maintain and trade with their own traditional units. Furthermore, although, for example, the firkin, kilderkin, strike, hogshead, tierce, pipe, butt, and puncheon were used in all the North American colonies, the actual amount to be associated with the same unit name might vary from colony to colony and even from locality to locality. A bushel of wheat in most parts of New Jersey weighed 32 pounds, but in Connecticut it usually weighed 28 pounds. In New York a whole set of Dutch measures were used among those with Dutch backgrounds, and in Florida and Louisiana around the time of the American Revolution, Spanish and French units, respectively, were often assumed.

The extent of the confusion, not to mention corrupt practices, generated by the plethora of units has been well described by Andro Linklater (2003):

> The settlers brought with them a gallimaufry of measures: Rhineland *Ruthkin* (rods) and Scots miles and Irish acres for measuring land; Scotch mutchins for beer; English flitches for bacon, fardels for cloth, and hattocks for grain; German *Globens* for flax, *Quentchens* for gold, and Zehnlingen for lambskins—in all, an estimated 100,000 units. For genuine frontier families who lived off what they raised, the variety of these local units presented no problem. Among themselves they used their own lengths and capacities. In western Pennsylvania these tended to be Rhineland measures, in Vincennes on the Wabash River they were French, in New York, Dutch. The only goods that needed to be measured by strangers were the ones they had to buy—flour, coffee, and sugar at the market, and cloth from peddlers—and no matter what their nationality, country dwellers knew that when they came to town, the scales were literally loaded in favor of the merchants who owned them. (p. 104)

George Washington, James Madison and Thomas Jefferson were among a minority in the new United States of America who recognized that unless appropriate action on the matter was taken quickly, then it would hard to achieve the large-scale change which was required. This was one case where arguments that local practices and rights should not be over-ruled by Congress needed to be challenged (Boyd, 1961). Jefferson succeeded with his decimalized currency, but the proposal to introduce a standardized, and decimalized, system of weights and measures, although permitted under the new constitution, was not adopted.

One is tempted to add nothing more than, "And, the rest is history"—but perhaps it would be in order for us to relate that in July 2014, as we were finalizing the proofs for this book, we noticed on the inside back cover of a 2014 "Mead Five-Star" notebook that we were using, a table which was headed "Handy Reference Guides." On the left side of the page was a 12 by 12 multiplication table, and printed next to this were tables of "dry measure," "linear measure," "liquid measure," and "miscellaneous measure." Without further comment we reproduce the relationships which were shown.

Dry Measure Table

2 pints (pt.)	=	1 quart (qt.)	4 pecks	=	1 bushel (bu)
8 quarts	=	1 peck (pk.)	1 cord	=	128 cu. Ft

Linear Measure Table

12 inches	=	1 foot	660 feet	=	1 furlong
3 feet	=	1 yard	320 rods	=	1 mile (5280 ft)
16½ feet	=	1 rod (5½ yds.)			

Liquid Measure Table

4 gills	=	1 pint (pt.)	31½ gallons	=	1 barrel (bbl.)
2 pints	=	1 quart (qt.)	2 barrels	=	1 hogshead (hhd.)
4 quarts	=	1 gallon (gal.)			

Miscellaneous Measure Table

12 units	=	1 dozen	1 fathom	=	6 feet
12 dozen	=	1 gross	1 knot	=	1 nautical mph
12 gross	=	1 great gross	1 nautical mi.	=	6080 feet
20 units	=	1 score	1 bu. potatoes	=	60 lbs.
1 hand	=	4 inches	1 barrel flour`	=	196 lbs.

From an educational point of view, Congress's decision to adopt a decimalized system of currency and then not to adopt a decimalized system of weights and measures raised interesting questions about what would happen with respect to the implemented school arithmetic curriculum in the United States of America. One might predict that there would be a greater emphasis on decimal fractions, but the traditional emphasis on money and on weights and measures would remain. In the next chapter, the effects on school arithmetic curricula during the period 1795–1810 will be examined.

References

Alder, K. (2002). *The measure of all things: The seven-year odyssey and hidden error that transformed the world*. New York, NY: The Free Press.

Boyd, J. P. (Ed.). (1953). *The papers of Thomas Jefferson, Vol. 7, March 1784 to February 1785*. Princeton, NJ: Princeton University Press.

Boyd, J. (1961). Report on weights and measures: Editorial note. In J. Boyd (Ed.), *The papers of Thomas Jefferson 16, November 1789 to July 1790* (pp. 602–617). Princeton, NJ: Princeton University Press.

Briggs, H. (1617). *Logarithmorum chilias prima*. London, UK: Author.

Burton, W. (1833). *The district school as it was, by one who went to it*. Boston, MA: Carter, Hendee and Company.

Cutter, W. R. (Ed.). (1914). *New England families genealogical and memorial: A record of achievements of her people in the making of the Commonwealth and the founding of a nation*. New York, NY: Lewis Historical Publishing Company.

Cocker, E. (1677). *Cocker's arithmetick: Being a plain and familiar method suitable to the meanest capacity for the full understanding of that incomparable art, as it is now taught by the ablest school-masters in city and country*. London, UK: John Hawkins.

Cocker, E. (1685). *Cocker's decimal arithmetick, …* London, UK: J. Richardson.

Denniss, J. (2012). *Figuring it out: Children's arithmetical manuscripts 1680–1880*. Oxford, UK: Huxley Scientific Press.

Dilworth, T. (1762). *The schoolmaster's assistant: Being a compendium of arithmetic both practical and theoretical* (11th ed.). London, UK: Henry Kent.

Earle, A. M. (1899). *Child life in colonial days*. New York, NY: The Macmillan Company.

Ellerton, N. F., & Clements, M. A. (2012). *Rewriting the history of school mathematics in North America 1607–1861*. New York, NY: Springer.

Ellerton, N. F., & Clements, M. A. (2014). *Abraham Lincoln's cyphering book and ten other extraordinary cyphering books*. New York, NY: Springer.

Greenwood, I. (1729). *Arithmetick vulgar and decimal, with the application thereof, to a variety of cases in trade and commerce*. Boston, MA: S. Kneeland & T. Green.

Grew, T. (1758). The description and use of the globes, celestial and terrestrial; with variety of examples for the learner's exercise: ... Germantown, PA: Christopher Sower.

Jefferson, T. (1785). *Notes on the establishment of a money unit and of a coinage for the United States*. Paris, France: Author. (The notes are reproduced in P. F. Ford (Ed.), *The works of Thomas Jefferson* (Vol. 4, pp. 297–313). New York, NY: G. P Putnam's Sons. This was published in 1904).

Kersey, J. (1683). *Mr Wingate's arithmetick, …* (8th ed.). London, UK: J. Williams.

Leybourn, W. (1690). *Cursus mathematicus*. London, UK: Author.

Linklater, A. (2003). *Measuring America: How the United States was shaped by the greatest land sale in history*. New York, NY: Plume.

Michael, I. (1993). The textbook as a commodity: Walkingame's *The Tutor's Assistant*. *Paradigm, 12*, 2–10.

Morris, R. (1782, January 15). Robert Morris to the President of Congress, January 15, 1782. In J. Boyd (Ed.), *The papers of Thomas Jefferson 16, March 1784 to February 1785* (pp. 160–169). Princeton, NJ: Princeton University Press.

Mouton, G. (1670). *Observationes diametrorum solis et lunae apparentium, meridianarumque aliquot altitudinum, cum tabula declinationum solis; Dissertatio de dierum naturalium inaequalitate,…* Lyons, France: Author.

Napier, J. (1614). *Mirifici logarithmorum canonis descriptio.* Edinburgh, UK: Andrew Hart.

Napier, J. (1619). *The wonderful canon of logarithms.* Edinburgh, UK: William Home Lizars, 1857 [English translation by Herschell Filipowski].

Pike, N. (1788). *A new and complete system of arithmetic, composed for the use of citizens of the United States.* Newbury-Port, MA: John Mycall.

Stafford, M. H. (1941). *A genealogy of the Kidder family: Comprising the descendants in the male line of Ensign James Kidder, 1626-1676, or Cambridge and Billerica in the colony of Massachusetts Bay.* Rutland, VT: Tuttle Pub. Co.

Stedall, J. (2012). *The history of mathematics: A very short introduction.* Oxford, UK: Oxford University Press.

Stevin, S. (1585). *De Thiende.* Leyden, The Netherlands: The University of Leyden.

Thomson, J. B. (1874). *Unification of weights and measures, the metric system: Its claims as an international standard of metrology.* New York, NY: Clark & Maynard.

Viète, F. (1579). *Canon mathematicus seu ad triangula cum appendicibus.* Paris, France: Jean Mettayer.

Wilkins, J. (1668). *Essay towards a real character and a philosophical language.* Held in archives of the University of Cambridge.

Wingate, E. (1624). *'L'usage de la règle de proportion en arithmétique.* Paris, France: Author.

Wingate, E. (1630). *Of natural and artificiall arithmetique.* London, United Kingdom: Author.

"Wright on measuring the meridian—Wright, Wren and Wilkins on an universal measure." (1805). *The Philosophical Magazine, 21,* 163–173.

Chapter 5
Decimal Fractions and Federal Money in School Mathematics in the United States of America 1787–1810

Abstract: In the preface to his 1788 textbook, *A New and Complete System of Arithmetic, Composed for the Use of the Citizens of the United States,* Nicolas Pike wrote that "as the United States are now an independent nation, it was judged that a system [of arithmetic] might be calculated more suitable to our meridian than those heretofore published" (p. vii). A glance at the contents of Pike's book reveals that it did have features not usually found in previous school arithmetics—for example, it included sections on logarithms, plain geometry, plain trigonometry, algebra and conic sections, and the sections on vulgar and decimal fractions appeared early in the book. But, Pike's book was rejected by many teachers as too difficult for most school children, and despite repeated calls for authors to write arithmetic textbooks which were specially designed for North American students, the most popular of the published arithmetics continued to emphasize the *abbaco*, commercially-oriented, arithmetic curriculum inherited from Europe. This chapter pays attention to immediate effects, during the period 1787–1810, that the introduction of a Federal, decimalized currency had on the author-intended arithmetic curriculum.

Keywords: Chauncey Lee; Daniel Adams; Decimal currency; George Washington; History of the dollar; Michael Walsh; Nathan Daboll; Nicolas Pike; Thomas Dilworth; Thomas Jefferson; U.S. Constitution; U.S. weights and measures

The Constitutional Background

Article 1, Section 8, Clause 5 of the U.S. Constitution gave Congress the power to fix the standard of weights and measures. That built on an earlier statement (1781), in the Articles of Confederation (Article IX, Section 4), that gave the central government "the sole and exclusive right and power of fixing the Standard of Weights and Measures throughout the United States." By 1786, however, the Articles were widely criticized and George Washington, Alexander Hamilton and James Madison were among those who moved to create a wholly new constitution. The original Articles of Confederation operated until early 1789 when the whole system was replaced by the new U.S. Constitution, which created a much stronger legal basis for national government (Bowen, 2007).

It is a historian's truism that the 1780s was a decade of optimism and opportunity so far as schooling in the United States was concerned. Sufficient, perhaps, to recall the general heightening of national consciousness, and strong, but uncoordinated, feeling that from that time onward everything that went on in the nation's education institutions should reflect the achievement of independence and its opportunity to create a model democracy.

There was much tension about how the new nation should govern itself, and it was not until May 25, 1787 that the Constitutional Convention met for the first time with the task of revising the Articles of Confederation. On September 17, 1787, the Convention's delegates voted to replace the Articles. But, even then the proposed Constitution needed to be ratified, clause by clause, by the states. That process ended on May 29, 1790, when Rhode Island agreed to ratify it (Bowen, 2007). Before the final ratification there was always a feeling, among politicians, that anything might happen. Of relevance to this book is the fact that, technically, school mathematics textbook authors would have felt it was presumptuous to incorporate the new decimal currency, or any decimalized scheme for weights and measures, until the U.S. Constitution had been clearly affirmed—only then could one be sure of what the national coinage and the national scheme for weights and measures would be.

Thomas Jefferson Fails to Achieve Reform in Weights and Measures, 1790–1795

Early in 1790, soon after his return from France to the United States, Thomas Jefferson was appointed Secretary of State and George Washington assigned him the task of drafting "a proper plan or plans for establishing uniformity in the currency, weights and measures of the United States" (Boyd, 1961, p. 604). Jefferson worked, under difficult circumstances, for months recalling, writing down, and polishing the radical scheme that he had originally conceived before he went to Paris. At this time he sought advice from and the support of numerous mathematicians and scientists on the issue.

Jefferson produced two drafts—one in April 1790 (see Boyd, 1961, pp. 624–628) and one in May (Boyd, 1961, pp. 628–648); finally, in July 1790, after much consultation, he produced his final report (Boyd, 1961, pp. 650–674). But despite George Washington's unequivocal support, appropriate legislation was never passed. Boyd (1961), whose analysis of the events is thorough and authoritative, has written:

> If ever a moment existed in the public mind for a general reformation … the summer of 1790 was assuredly that moment. Not only had the public become accustomed in the preceding year to serious discussions of radical proposals of reform in weights and measures, but no open opposition or even complacency … seems to have voiced. … With an apparently receptive public and with the powerful support of Washington, Hamilton and Madison, Jefferson's reform, it seems, could scarcely have failed of adoption. But such a fortunate conjunction of circumstances never appeared again, and thus was lost the opportunity that many viewed as one for setting an example to the world. (pp. 614–615)

And so, the United States missed its chance to do something very similar to what France would do a few years later. Congress ordered that several calculation errors found in the final report be corrected. But, when the "correct" version of the final draft was ready, the politics had changed.

In fact, Jefferson's final report offered two alternative schemes for a national system of weights and measures. The first scheme was the more conservative: it did take into account what had been done with currency, and it did link length, area, volume, capacity, weights, and coinage—but it did not have decimal relationships built into it.

Jefferson's strong preference was for the second system he put forward in the following way:

> But if it be thought that either now or at any future time, the citizens of the U.S. may be induced to undertake a thorough reformation of their whole system of measures, weights and coins, reducing every branch to the same decimal ratio, already established in their coins, and thus bringing the calculation of the

principal affairs of life within the arithmetic of every man who can multiply and divide plain numbers, greater changes will be necessary. (Quoted in Boyd, 1961, p. 640)

Then followed Jefferson's plan for a fully decimalized system—one which had been incubating in his fertile brain for many years. That system was based on a unit of time derived from a second's rod (see Boyd, 1961, pp. 640–648). This second plan was elegant, holistic, and absolutely radical. If approved, it would have preceded the French metric system as the world's first fully decimalized and coordinated system of weights and measures.

The historian will never stop wondering why Jefferson presented it as the second of the two proposals in his final report. Boyd (1961) suggested that it could have been a strategy, whereby once members of Congress had seen how tame the first proposal was in comparison with the second, members would vote for the second. Another possibility is that Jefferson believed that although his second proposal was the one he wanted, acting on his first proposal would be much better than not acting at all. But, as it turned out, the delays in 1790 and 1791 allowed enough respite for the opponents of Jefferson's scheme of weights and measures to gather their forces together, and the result would be that the scheme was lost. Land between the Appalachians and the Mississippi River, and beyond, opened up for sale—everyone knew that the nation desperately needed the money from the sale of the land, and there were plenty of interested buyers. However, before selling could begin, the land needed to be surveyed. Therefore, many reasoned, it would not be sensible to delay sales by forcing surveyors to learn to use an entirely new system of weights and measures (Linklater, 2003). And so, from a mathematics education perspective, nothing would be achieved.

In Chapter 1 (see, especially, Figure 1.3) we presented a lag-time theoretical framework, in relation to mathematics curriculum development, which encompassed intersecting domains of research mathematics, service mathematics and mathematics education embedded within ethnomathematical contexts which embraced politics, family, community, and work. From that perspective, one might predict that the United States of America's introduction of a Federal decimal currency would have resulted in decimal fractions becoming a more vital component of implemented school arithmetic curricula. One might also suppose that the pace of acceptance of decimal fractions in curricula would have been increased if the move around 1790 for the introduction of a national decimalized system of weights and measures had been adopted. Later in this chapter U.S. cyphering-book evidence for the periods 1792–1800 and 1800–1810 will be analyzed, the aim being to find evidence for, or against, the lag-time theory with respect to the place of decimals in school curricula.

But first it will be useful to consider how, in the late 1780s and throughout the 1790s, mathematics textbook authors responded to the introduction of a Federal decimalized currency, and to moves to introduce a national coordinated system of weights and measures.

Nicolas Pike and his *New and Complete System of Arithmetic*

Prior to the Revolutionary War (1775–1783), most of the arithmetic textbooks used in the North American British colonies had been written by British authors (e.g., Cocker, 1685, 1719, 1738; Dilworth, 1762). Immediately after the Revolutionary War there was a surge of activity in American publishing for schools and colleges (Monroe, 1917). The first major school arithmetic written by a North American author was Pike's (1788) portentous 512-page *A New and Complete System of Arithmetic Composed for Use of Citizens of the United*

States. Pike (1743–1819), a native of New Hampshire, had graduated from Harvard College in 1766, and had then taught mathematics for many years.

As well as Pike's (1788) *A New and Complete System of Arithmetic,* which first appeared in 1788, a 1793 *Abridgement of the New and Complete System of Arithmetick* would also be published. The *Abridgement,* which had 371 pages, was significantly smaller than *A New and Complete System of Arithmetic.* It was particularly aimed at schools, whereas the larger *New and Complete System of Arithmetic* had been aimed at both schools and colleges. The publisher's preface to the *Abridgement* stated that the original *Arithmetic* was "now used as a classical book in all the New England universities," and excelled "everything of the kind on this content." The author had "reason to hope that this abridgement will not be less esteemed as a school book" (Pike, 1793, p. ix).

"Old Pike," as Pike's (1788) original *Arithmetic* would come to be known, would go through six editions between 1788 and 1843 (Albree, 2002; Karpinksi, 1980; Thomas & Andrews, 1809). In 1788 it sold for about $2.50—at that time a substantial price for a textbook, and one which placed it out of the reach of most pupils and teachers (Monroe, 1917). Besides arithmetic proper, it introduced algebra, geometry, trigonometry, and conic sections. Applications of arithmetic were made to problems associated with mechanics, gravity, pendulums, mechanical powers, and to astronomical problems requiring calculations of the moon's age, the times of its phases, and the date of Easter. But these additional topics were probably deemed to be irrelevant by schoolmasters, and in the 1793 *Abridgement,* and in later editions of that *Abridgment,* the more advanced topics were dropped—leaving Pike's text as strikingly similar to the British arithmetics that he had hoped to supplant.

Pike's Ambitions Foiled by the Slow Process of Constitutional Politics

It is almost certain that Nicolas Pike was very frustrated by the slow process of constitutional politics. It seems that he had his original textbook ready for publication at least three years before it was actually published—for in 1785 Benjamin West, an astronomer based in Rhode Island, declared that he had perused Pike's book and found it to be suitable for use in North American schools and colleges. In West's opinion, arithmetic textbooks by well-known British authors "Wingate, Hutton, Ward, Hill, and many other authors" were "inadequate and defective" when compared with Pike's *Arithmetic*—see the full text of West's recommendation on p. iv of Pike (1788). Pike was also able to secure glowing recommendations of his book from the presidents of Yale, Harvard, and Dartmouth Colleges, from several of the professors of mathematics at those colleges, and from Governor Bowdoin of Massachusetts. In June 1788, even President George Washington, in acknowledging Pike's written request for a recommendation, offered guarded congratulations:

> I hope and trust that the work [i.e., Pike's text on arithmetic] will prove not less profitable than reputable to yourself. It seems to have been conceded, on all hands, that such a system was much wanted. Its merits being established by the approbation of competent judges, I flatter myself that the idea of its being an American production, and the first of the kind which has appeared, will induce every patriotic and liberal character to give it all the countenance and patronage in his power. In all events, you may rest assured that, as no person takes more interest in the encouragement of American genius, so no one will be more highly gratified with the success of your ingenious, arduous and useful undertaking. (George Washington to Nicolas Pike, June 20, 1788)

Pike's confidence in the value of his work was evidenced by the fact that before his book was published he registered as author in Pennsylvania, South Carolina, Massachusetts, and New York, such registration serving as copyright notice.

But during the time he was preparing his text Pike was hamstrung by the slow pace of constitutional development. Much of the subject of arithmetic concerned weights and measures, and Pike naturally would have wanted to know final details of any national systems of coinage and weights and measures. He held the date of publication back as long as he could. But in the end he could wait no longer. In the Preface to his 1788 textbook Pike claimed that existing arithmetic textbooks were defective and, in any case, "as the United States are now an independent nation, it was judged that a system might be calculated more suitable to our meridian than those heretofore published" (p. vii). Pike noted, in his Preface, that "the Federal coin, being purely decimal, most naturally falls in after decimal fractions" (p. vii). Consistent with this statement was the fact that the section on "Federal Money" (pp. 96–100) was placed immediately after the section on decimal fractions" (pp. 85–96). However, the section on Federal money occupied just four pages, and much of that was a statement of Congress's decision (on "8th of August 1786") to adopt a decimal system of coinage which would feature the mill (1000th of a dollar), the cent (100th of a dollar), the dime (10th of a dollar), the dollar, and the eagle (10 dollars) (pp. 96–97). After little more than a page showing examples of how Federal money could be added, subtracted, multiplied and divided (pp. 97–98), Pike then moved on to other topics, and throughout the book the examples involving money were expressed in pounds, shillings and pence (the only exception being on page 155, when the topic of "Exchange" was being considered).

So far as weights and measures were concerned, throughout the book Pike used all the traditional British units for avoirdupois, troy and apothecaries weight, for length, area and volume, for capacity, and for time. Most of the text was devoted to narrow forms of traditional arithmetic. The book began with the rules for the elementary operations for integers, together with many examples worked out in detail. Then followed sections on vulgar fractions, decimal fractions, rules for exchanging currency, tricks for rapid computation, extraction of square roots, computation of interest, commissions, annuities, the volumes of particular solids, arithmetical and geometrical progressions, permutations and combinations, and topics from elementary mechanics. The book was, essentially, a detailed compendium of useful techniques and formulas, with examples completely worked out, in a wide diversity of practical applications. There were very few, if any, formal proofs. The formulae presented for the slightly more advanced topics appeared without any details being given with respect to their origins. None of the problems involved the new American money.

In his choice and ordering of content and his methods of handling the various topics, Pike leant heavily on school arithmetics written and published in England which had been widely used in the American colonies—especially those written by John Bonnycastle, Edward Cocker, Thomas Dilworth, and John Ward. Like the authors of those arithmetics, Pike claimed that his book was "practical." For example, they all had rules for the "reduction of coin." Naturally, those written in England assumed that English "pounds, shillings, pence" would be solely used in the schools. This assumption was translated into the American colonies. Thus, although Pike's (1788) *Arithmetic* devoted 28 pages (pages 96–123) to currency conversion, only four of those (pages 96–98) were concerned with the new Federal currency. The remaining pages stated and illustrated rules for transferring between the various currencies that had been used in the old colonies. Pike included sections on converting New Hampshire, Massachusetts, Rhode Island, Connecticut, and Virginia currency to Federal money, to New York and North Carolina currency, to Pennsylvania, New Jersey, Delaware and Maryland currency, to Irish money, to Canadian and Nova Scotia currency, to *livres tournois*, and to Spanish milled dollars. He gave specific rules for reducing Federal money to New England and Virginia currency, etc.

Other than in the brief section on Federal money, Pike's (1788) text reflected a viewpoint that much of the time spent in schools on cyphering should be dedicated to rules and calculations related to conversion of units. However, he expected students to follow rules on monetary conversions suited to old-society currencies and measures and offered hardly any examples of how the new Federal currency could be applied in farming, trade and business transactions.

A page from Pike's (1788) *Arithmetic* illustrating his dilemma. Figure 5.1 is from page 131 of Pike's original *Arithmetic*. It shows the following 12 examples that related to the "single rule of three"— which was the title of the relevant section in the book:

20. *A* owes *B* £3475, but *B* compounds with him for 13*s* 4*d* on the pound; pray, what must he receive for his debt?

21. If the distance from Newbury Port to York be 31 miles, I demand how many times a wheel, whose circumference is 15½ feet, will turn round in performing the journey?

22. Bought 9 chests of tea, each weighing 3 *C,* 2 qrs, 21 lb at £4 9*s* per cwt, what came they to?

23. What will 37½ gross of buttons at 9½*d* per dozen, come to?

24. A farm, containing 125 *A,* 3 *R,* 27 *P* is rented at £3 9*s* per acre. What is the yearly rent of that farm?

25. If a ship cost £537, what are 3/8 of her worth?

26. If 7/16 of a ship cost £349, what is the whole worth?

27. Bought, a cask of wine at 4/7 per gallon, for 125 dollars, how much did it contain?

28. What comes the insurance of £537 15*s* 10 at 4½ per centum?

29. What come the commissions of £785 to at 3½ guineas per cent?

30. A merchant bought 9 packages of cloth. At 3 guineas for 7 yards; each package contained 8 parcels, each parcel, 12 pieces, and each piece 20 yards. What came the whole to? And what per yard?

31. A merchant bought 49 tuns of wine for £273; freights cost £27, duties £12, other charges £15, and he would gain £55 10*s* by the bargain. What must I give him for 23 tuns?

Pike's (1788) textbook included a 12-page section (pages 85–96) specifically dedicated to the concept of a decimal fraction and to rules for adding, subtracting, multiplying and dividing decimal fractions. An impressive numeration table was shown, on page 85, depicting numbers from hundreds of millions down to 1, and then down to decimal fractions for numbers like 100 millionths. This section came immediately after the section on vulgar fractions and before exercises on business applications.

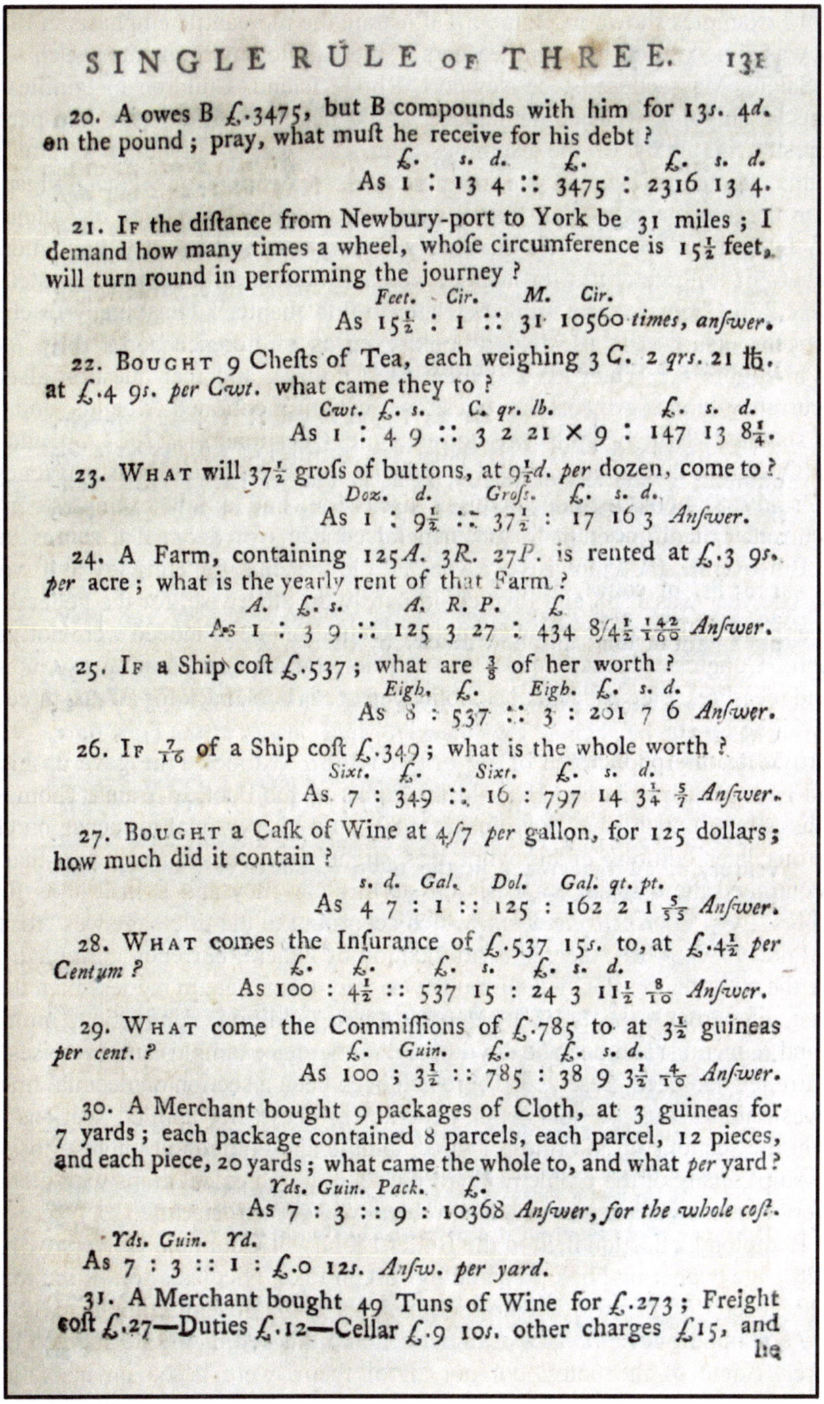

SINGLE RULE of THREE. 131

20. A owes B £.3475, but B compounds with him for 13s. 4d. on the pound ; pray, what must he receive for his debt ?

£.   s. d.   £.   £.   s. d.
As 1 : 13 4 :: 3475 : 2316 13 4.

21. If the distance from Newbury-port to York be 31 miles ; I demand how many times a wheel, whose circumference is 15½ feet, will turn round in performing the journey ?

Feet.   Cir.   M. Cir.
As 15½ : 1 :: 31 10560 *times, answer.*

22. Bought 9 Chests of Tea, each weighing 3 C. 2 qrs. 21 ℔. at £.4 9s. *per Cwt.* what came they to ?

Cwt. £. s.   C. qr. lb.   £. s. d.
As 1 : 4 9 :: 3 2 21 × 9 : 147 13 8¼.

23. What will 37½ gross of buttons, at 9½d. *per dozen,* come to ?

Doz. d.   Grofs. £. s. d.
As 1 : 9½ :: 37½ : 17 16 3 *Answer.*

24. A Farm, containing 125A. 3R. 27P. is rented at £.3 9s. *per* acre ; what is the yearly rent of that Farm ?

A. £. s.   A. R. P.   £.
As 1 : 3 9 :: 125 3 27 : 434 8/4½ 142/160 *Answer.*

25. If a Ship cost £.537 ; what are ⅜ of her worth ?

Eigh. £.   Eigh. £. s. d.
As 8 : 537 :: 3 : 201 7 6 *Answer.*

26. If 7/16 of a Ship cost £.349 ; what is the whole worth ?

Sixt. £.   Sixt. £. s. d.
As 7 : 349 :: 16 : 797 14 3¼ 5/7 *Answer.*

27. Bought a Cask of Wine at 4/7 *per* gallon, for 125 dollars ; how much did it contain ?

s. d. Gal. Dol. Gal. qt. pt.
As 4 7 : 1 :: 125 : 162 2 1 5/55 *Answer.*

28. What comes the Insurance of £.537 15s. to, at £.4½ per Centum ?

£.   £.   £. s. £. s. d.
As 100 : 4½ :: 537 15 : 24 3 11½ 8/10 *Answer.*

29. What come the Commissions of £.785 to at 3½ guineas per cent. ?

£. Guin. £.   £. s. d.
As 100 ; 3½ :: 785 : 38 9 3½ 4/10 *Answer.*

30. A Merchant bought 9 packages of Cloth, at 3 guineas for 7 yards ; each package contained 8 parcels, each parcel, 12 pieces, and each piece, 20 yards ; what came the whole to, and what *per* yard ?

Yds. Guin. Pack. £.
As 7 : 3 :: 9 : 10368 *Answer, for the whole cost.*

Yds. Guin. Yd.
As 7 : 3 :: 1 : £.0 12s. *Answ. per yard.*

31. A Merchant bought 49 Tuns of Wine for £.273 ; Freight cost £.27—Duties £.12—Cellar £.9 10s. other charges £15, and he

Figure 5.1. Page 131 from Nicolas Pike's (1788) *Arithmetic.*

The 12 examples shown in Figure 5.1 illustrate the mercantile emphasis in Pike's book. The book was aimed at teachers and learners in mercantile environments—such as might be found in Salem, Massachusetts, or Newport, Rhode Island. Children in families in which investments in shipping were commonplace would have identified with, and perhaps have been interested in, many of the examples. But, the learner attending a rural school in Pennsylvania would have found the examples less relevant. Notice that vulgar fractions appeared in the statements of 6 of the 12 examples. Pike had a section on vulgar fractions quite early in his book (pp. 70–85), but analyses of cyphering-book data have indicated that most teachers of arithmetic did not include vulgar fractions in their implemented curricula (see Table 2.2 in Chapter 2 and Table 5.1 later in this chapter). Thus, many teachers would not have been persuaded to use Pike's book, even as a reference book. Only 1 of the 12 examples in Figure 5.1 (Question 27) mentioned dollars, and that question also involved sterling currency units. All of the questions which mentioned weights and measures employed standard "British" units (see, for example, Questions 21, 22, 24, 30, and 31). Two questions (Questions 29 and 30) referred to guineas, a quintessential British currency unit.

Pike's (1788) book included a brief discussion, and a few examples, on the new decimal currency. It did attempt to link general concepts of decimal fractions to the new currency. But overall, the emphasis was on the old systems for currency and weights and measures. The year was 1788 and Pike could hardly do otherwise, for the Federal Mint was not yet open, and dollars and cents were not in circulation—and indeed were not yet minted. Furthermore, Congress had not made any decisions about inaugurating a new system of weights and measures. Pike had held the publication of his book back for at least three years, but he could wait no longer.

Shortly after the publication of the original 1788 textbook Pike gave up his teaching career and became a magistrate. He sold the rights to his book to Isaiah Thomas, a well-known Massachusetts publisher (Karpinski, 1980), and his personal influence on changes to the numerous later editions of his work was slight. One consequence was that the later editions continued the original emphasis on sterling currency and British units for weights and measures. Even in his *Abridgement*, which according to the title page was "for the use of schools" (Pike, 1793, p. iii), there was no mention of Federal currency until page 101—and even there the treatment of decimal currency was very brief, occupying less than three pages. Before that, numerous aspects of sterling currency (addition, subtraction, multiplication, division, and reduction) had been dealt with, and numerous examples and exercises involving sterling currency had been presented. There had even been a section on decimal fractions, but nothing was said in that section about Federal currency. In "applied sections" (like for example, in the sections on the rules of three, simple and compound interest, loss and gain, and fellowship), many of the problems that Pike set required calculations with money, and in almost every case the currency in which a problem was set was sterling.

The Plimpton Collection held in the Butler Library at Columbia University, New York, holds a 128-page cyphering book prepared by Joseph Pike, Nicolas's son (Plimpton MS 511, 1799–1800, Joseph Somersworth Pike). Within this book, Joseph completed 128 pages of elementary and middle-level *abbaco* arithmetic, and in his setting out he adopted IRCEE and PCA genres. Some of the notes, but not all of them, were based on his father's 1788 textbook. That Joseph prepared his own cyphering book was testimony to how society believed that the preparation of one's own handwritten cyphering book was of great value from an education perspective.

George H. Martin's (1897) Case Against Nicolas Pike

Many critics of Pike's (1788) *Arithmetic* (e.g., Cobb, 1835) pointed out that the numerous rules Pike gave were incomprehensible to most school students. In 1897, George H. Martin, in tracing the evolution of the Massachusetts public school system, launched a scathing attack on what he perceived to be the enduring negative effects of Pike's (1788) *Arithmetic* on schooling in the United States. Martin (1897) wrote:

> The labor involved in the computation of ordinary business transactions at this period is almost appalling. The money units were the English; two pages only are given to Federal money, as it was called, which the Congress had just established but which had not come into general use. Nine kinds of currency were in use in commercial transactions, and the students of this arithmetic were taught to express each in terms of the others, making 72 distinct rules to be learned and applied. (p. 102)

An examination of passages in Pike (1788) suggests that some of these criticisms were warranted. For example, for the topic "Practice," which Pike described as "an easy and concise method of working most questions which occur in trade and business," the learner was required to memorize a page of tables of aliquot parts of pounds and shillings, of hundredweights and tons, and a table of per cents of the pounds in shillings and pence. These tables included more than 100 relationships, and there were 34 cases, each with a rule. Thus, for example, "Case 12" stated:

> When the price is shillings, pence and farthings, and not an even part of a pound, multiply the given quantity by the shillings in the price of one yard, etc., and take parts of parts from the quantity for the pence, etc., then add them together, and their sum will be the answer in shillings, etc. *Or,* you may let the given quantity stand as pounds per yard, etc., then draw a line underneath, and take parts of parts therefrom; which add together, and their sum will be the answer. (Pike, 1788, p. 169)

After that statement Pike advised the learner "to work the following examples both ways by which means he will be able to discover the most concise method of performing such questions in business, as may fall under this case" (p. 169).

Under the topic "Tare and Trett" the following rule was given as "Case 4," which was meant to relate to the situation "when tare, trett and cloff are allowed":

> Deduct the tare and trett … divide the suttle by 168, and the quotient will be the cloff, which subtract from the suttle, and the remainder will be the neat. (Pike, 1788, p. 194)

Martin (1897) maintained that the book "gave tone to all the arithmetic of the district-school period" (p. 104), and was "responsible for that excessive devotion to arithmetic which has of late been the subject of just complaint" (p. 104). He stated that Pike's text had "an almost endless elaboration of cases and prescription of rules" (p. 104). There were 14 rules under simple multiplication, and in all there were 362 rules in the book. According to Martin, except in an occasional footnote no hint of a reason for any rule was given, and often there were many very difficult problems. Martin went on to say that most district-school pupils, and especially the girls, did not do arithmetic beyond the four fundamental operations and a brief consideration of vulgar fractions. He maintained that if they ever reached the celebrated "rule of three," they would be regarded as "mathematicians," and if, later, someone went right through "Old Pike" they were likely to be regarded as a "prodigy" and be eagerly sought after for teaching appointments.

Cajori's (1907) defense of Pike against Martin's accusations. Florian Cajori, a mathematics and physics professor and respected historian of mathematics and mathematics education, reacted sharply to Martin's (1897) criticisms of "Old Pike." Cajori (1907) argued that Pike's emphasis on local, non-Federal currencies was appropriate because those were the kinds of calculations people needed to know how to do if they were to survive with dignity in everyday life at a time when the different currencies of the American colonies were causing much confusion. Cajori also emphasized that Pike did cause to have published, in 1793, his "*Abridgement* for the Use of Schools" (p. 217). Then Cajori (1907), referring directly to Martin's (1897) criticisms of Pike, wrote:

> To us, this [Martin's] condemnation of Pike seems wholly unjust. It is unmerited, even if we admit that Pike was in no sense a reformer among arithmetical authors. Most of the evils in question have a far remoter origin than the time of Pike. Our author is fully up to the standard of English authors to that date. He can no more be blamed by us for giving the aliquot parts of pounds and shillings, for stating rules for "tare and trett," for discussing the "reduction of coins," than the future historian can blame works of the present time for treating of such atrocious relations as that 3 ft. = 1 yd., 5½ yds. = 1 rd., 30¼ sq. yds. = 1 sq. rd., etc. So long as this free and independent people chooses to be tied down to such relics of barbarism, the arithmetician cannot do otherwise than supply the means of acquiring the precious knowledge. (p. 218)

Cajori (1907) then warmed to his defense of Pike:

> At the beginning of the nineteenth century there were three "great arithmeticians" in the United States: Nicholas Pike, Daniel Adams, and Nathan Daboll. The arithmetics of Adams (1801) and Daboll (1800) paid more attention than that of Pike to Federal money. ... As a consequence of the general use, for over a century, of Dilworth in American schools, pounds, shillings, and pence were classical, and dollars and cents vulgar for several succeeding generations. "I would not give a penny for it," was genteel; "I would not give a cent for it," was plebian. (p. 218)

Cajori's (1907) defense of Pike, then, went something like this: Nicolas Pike, and no doubt his publisher, John Mycall, recognized that in 1788 the most likely users of his text would be students at higher educational institutions like Harvard and Yale, and students in the higher academies. Of course, the author and the publisher would be prepared to maintain that the book was "suitable for schools," and so it was (if, by schools, we included "academies that prepared students for higher study"). Cajori might also have added that an "abridged version" of Pike's original arithmetic, which was more suitable for common schools, appeared in 1793, and that this abridged version continued to be published until the 1830s (Karpinksi, 1980).

However, it could be argued that the fact that the 1793 *Abridgment* needed to be published testified to the unsuitability of the original (1788) *Arithmetic* for schools—despite the fact that all four "recommendations," written by eminent citizens of Boston and printed in the front of the 1788 edition, had indicated that the text would be very useful in *all* schools. For example, the recommendation by Benjamin West stated that the 1788 edition would be read "by great advantage by students of every class, from the lowest school to the university" (p. 4). History would show that not everyone agreed with that assessment. Monroe (1917), for example, in his history of the development of arithmetic as a school subject in the United States, stated that Pike's (1788) *Arithmetic* was "not a text for young pupils" (p. 18).

Cajori maintained it was unfair of Martin to have expected Pike to be a century ahead of his time so far as his thinking about the needs of average school students was concerned. Pike had been a practicing teacher, a product of a system transported from England by which a textbook was expected to state rules which students would copy, and attempt to remember. That is what teachers expected "Old Pike" to do, and that is what it did. For Cajori, Pike's (1788) summary of relationships between local currencies was meritorious—indeed, such was the detail provided that the book became an authoritative reference for numismatic experts (see, e.g., Mehl, 1933).

Final Comments on the Influence of Nicolas Pike

Cajori (1907) believed that it was not an arithmetic textbook author's task to set out to change the way people used currencies within society. Rather, it was his, or her, job to make sure that students learned to cope, arithmetically, with the ways currencies were being used on a daily basis. Furthermore, Pike's (1788) emphases on rules was in line with the "best thinking" on teaching and learning. The issue is whether Pike should have accepted the existing education settings of his day, and proceeded cautiously, taking into account his contextual constraints; or whether he should, as a person acting at a pivotal period of history, have provided leadership by seizing the moment and attempting to achieve fundamental, even radical, change, in the teaching and learning of arithmetic.

Pike (1788) expected that his book would be a landmark publication, and so did all the notable personalities who allowed their names to be used, and provided supporting statements, in the acknowledgements section at the front the book. Pike wanted his book to be the first English-language arithmetic text written by a U.S. citizen to be widely used in the schools of the new nation. One could argue that in those circumstances he had the responsibility to be brave, to set a new tone, to break away from the colonialist fetters that, he believed, had strangled teaching and learning in the schools before the Revolutionary War. But, he failed to seize the moment. Furthermore, the abridged versions of Pike's arithmetic, "for schools," were little better than the original 1788 text.

Was it unreasonable to have expected Pike to see beyond the horizons surrounding his world and context in the 1780s? That question raises intriguing questions of historiography. What principles can historians look to if they want to generate faithful yet historical accounts of events and penetrating, insightful interpretations of those events? Under what circumstances is it fair to criticize a writer for "silence" about ideas and practices of which he was totally unaware, or only dimly aware? Those kinds of questions are fiercely contested within the world of academic history today (see e.g., Macintyre & Clark, 2004; Windschuttle, 1996).

If Pike can be defended on the grounds that, in 1788, the Federal Mint had not yet been opened, and there were no actual coins (eagles, dollars, dimes, and cents) to justify the inclusion of more than a brief mention of Federal currency in his book, that excuse was no longer justified when the *Abridgement* for schools first appeared in 1793. And, it certainly was not justified in 1797 when, in the second edition of Pike's original textbook (Pike 1797), there was little more than a nod in the direction of Federal currency, with most money tasks being set in the context of sterling currency. By 1797 Pike, had moved out of teaching to become a magistrate, and he would have argued that he was not responsible for the content of the 1797 text because he no longer held the copyright—that was now held by Isaiah Thomas, the publisher. In fact, Isaiah Thomas wrote the Preface to the 1797 edition. But, the title page proclaimed that the book was written "by Nicolas Pike, Esq, Member of the American Academy of Arts and Science," and Pike was permitting his name to be used with the text. So far as the lack of appropriate emphasis on Federal currency was concerned, Pike had a case to answer.

Reluctance of Other Textbook Authors to Decimalize Money 1788–1798

Gough's and Workman's Texts

It is hardly surprising that, after the Constitution had been approved by Congress, confirming, among other things, the earlier decision to decimalize U.S. Federal currency, there was an increase in the number of arithmetics written by U.S. citizens. In 1788, a 370-page book written by John Gough, of Ireland, was published in Philadelphia. The title of this book was *A Treatise of Arithmetic in Theory and Practice Containing Everything Important in the Study of Abstract and Applicant Numbers, Adapted to the Commerce of Great Britain and Ireland*. Despite the fact that Benjamin Workman, a U.S. teacher, claimed that he had added "many valuable and amendments more particularly fitting to the work for the improvement of the American youth," the book was hardly different from Gough's text written for Ireland. Except for a section on "Exchange," all of the questions involving money referred to pounds, shillings, pence and farthings. There was no mention of the fact that Congress had approved a system of decimal arithmetic for the United States of America and, indeed, none of the word problems in the text had been revised so that they would be set in North American contexts.

Such was the lack of success of this blatant attempt to take advantage of the North American educational scene, that one year later Workman (1789) caused to have published, also in Philadelphia—and through the same publisher who had published Gough's text in 1788—a 224-page textbook entitled *The American Accountant or Schoolmasters' New Assistant*. In his Preface, Workman said that he was dropping the theoretical components of Gough's book, because he wanted to "furnish the scholar, at a cheap rate, with a complete system of practical arithmetic." This book had greater success, with second and third editions being printed in 1793 and 1796, but even in those new editions there were hardly any references to the new Federal currency. Workman (1789) informed his readers that "in England, Ireland, and America, accounts are kept in pounds, shillings and pence" (p. 33), and almost all money examples were consistent with that backward-looking statement. Workman (1789) did have a brief section on "decimal fractions" (pp. 93–100), but nowhere in that section was there any mention of the new Federal currency. In later editions, Workman did refer, briefly, to "Federal Money," which he said had been approved by "Acts of Congress 1792 and 1793" (see, e.g., Workman, 1793, p. 34). He also stated that "10 mills make a cent, 10 cents make a dime, 10 dimes make a dollar [for which he used the symbol *D*] and 10 dollars make an eagle" (p. 34). But that was all he had to say about Federal money, and no examples were given, or exercises set in relation to the new currency.

Consider Sterry and John Sterry's 1790 and 1795 textbooks

The 1790 textbook by the Sterrys. In 1790 the brothers Consider and John Sterry caused to be published, in Providence Rhode Island, a 388-page book titled *The American Youth: Being a New and Complete Course of Introductory Mathematics, Designed for the Use of Private Students*. What was most notable about this text was that, like Nicolas Pike's (1788) text, it included a section on algebra—indeed, 147 pages were devoted to algebra (whereas Pike had devoted only 39 pages to algebra.). The Sterrys were private teachers, outside of college circles, and so it was not to be expected that they would include such an extensive, and mathematically ambitious, section on algebra (Simons, 1924). The extent of the algebra covered was such that the text provided the widest coverage on algebra of any textbook written by North American citizens in the eighteenth century.

In their Preface, Sterry and Sterry (1790) maintained that existing mathematics textbooks were "not adapted to the capacity of young and tender minds" (p. v), mainly because the authors had paid too much attention to "close and refined reasoning" and not enough to "simplicity, plainness and brevity" (p. v). The Sterrys probably had Nicolas Pike's book in mind when they commented that some books were "so prolix and voluminous, as even to discourage a learner at the sight of their works" (p. v).

The Sterrys paid considerable attention to the new Federal decimal currency, showing how operations on sums of money could be carried out by decimal operations. Curiously, though, that was done before the 32-page section on decimal fractions.

The Sterry's book did not go to a second edition, and so it is reasonable to assume that not many teachers or students ever used it. Probably, the section on algebra was off-putting for many. Furthermore, it was published in 1790, two years before the United States Mint was established—there were no decimal coins in circulation, and everyone was still using sterling currency, or Spanish dollars. Neither the teachers nor the students wanted to spend their time learning to calculate with a system of currency that was not operational at that time.

So far as weights and measures were concerned, the Sterrys offered a traditional coverage of all the different kinds of measures. In the absence of any progress in Congress on decimalizing weights and measures, who could blame the Sterrys for the approach they took?

The 1795 textbook by the Sterrys. The Sterry's (1795) much smaller text, *A Complete Exercise Book in Arithmetic, Designed for the Use of Schools in the United States,* had 121 pages. There was no algebra, and a larger treatment of decimal currency was provided than in the earlier book—which was appropriate, given that by 1795 the U.S. Mint had been open for three years, and people were now expected to learn to deal with the new coinage. Like the 1790 textbook, the Sterry's 1995 book did not go to a second edition.

In their Preface the Sterrys emphasized that this new, smaller book was aimed at schools. The section on compound operations offered specific instructions with respect to decimal currency:

> In addition of money of the United States, add the numbers as in simple addition, and separate with a point, as many figures on the right hand, as are equal to the greatest number in the inferior denominations given in the question; then

decimate those on the right hand, beginning at the point, and call the first figure dimes, the second cents, third mills, etc. (p. 20)

[For subtraction of Federal money] In the money of the United States, point and decimate as in addition. (p. 25)

[For multiplication] If one of the factors is dollars, cents, &c, multiply and separate on the right hand of the product as many figures as there are cents, mills, &c, and decimate as before in addition. (p. 28)

[For division] In division of money of the United States, divide as in simple division, and separate with a point as many figures on the right of the quotient as there are contained in the inferior denominations in the dividend; then decimate as before taught. (p. 31)

[For Reduction] In reducing dollars, dimes, cents, and mills, to mills, the given numbers wrote as one in a line will be reduced as required. 46 dol 2 d 6 c 3 m reduced to mills is 46263 mills. (p. 35)

The language used in these instructions would have been formidable for many students. Furthermore, the rules were given well before the section in the book on decimal fractions— which occupied pages 49 through 58.

A Complete Exercise Book in Arithmetic did not begin with a numeration table. The Sterrys began their book with the following "definitions":

Arithmetic is the art of composition by numeral figures, called *digits* [original emphasis]. Which are considered either integral or fractional, and therefore vulgar or decimal.

Vulgar arithmetic contemplates these digits integrally or dividedly.

Decimal arithmetic considers those digits in a decimal ratio of those parts to unity.

The digits made use of are these, 1, 2, 3, 4, 5, 6, 7, 8, 9, and for convenience in computation is added the cypher, 0.

All arithmetical operations are performed by addition, subtraction, multiplication and division. (p. 5)

Sterry and Sterry (1795) were obviously attempting to sum up the achievement of those who developed the Hindu-Arabic numeration system, and of those who had extended it to include vulgar fractions and decimal fractions. This introduction to numbers should already have been met by students before they began to prepare a cyphering book, but the Sterrys' introduction went further than most when they made explicit reference to vulgar and decimal fractions.

In keeping with this early mention of vulgar and decimal fractions, Sterry and Sterry (1795) provided more than 12 pages on vulgar fractions quite early in their book (pp. 37–49). They immediately followed their section on vulgar fractions with a 10-page section on decimal fractions (pp. 49–58). Thus, 22 pages of their 120-page book were dedicated to vulgar and decimal fractions, and the 22 pages came quite early in the book. This approach was unusual for the 1790s.

Furthermore, the 22 pages were placed in the book before more advanced topics like the rules of three were introduced—in fact, the Sterrys followed their introduction to decimal fractions with the rules of three, the rules of practice, tare and tret, simple and compound interest, commission, brokerage, rebate or discount, barter, loss and gain, fellowship and alligation. That order of the topics was according to the *abbaco* traditional, except that in this book vulgar and decimal fractions had already been dealt with, and therefore students could use the fractions in the traditional topics. Thus, the Sterrys were able to ask students to tackle exercises such as:

- A goldsmith sold a tankard for 29 dol. 97 cts. It weighed 270 oz. What is that per ounce? (p. 60)
- If 20 bu. of grain at 50 cts. per bushel will pay a debt, how much at 2 d 60 ct will pay the same? (p. 63)
- Two partners, *A* and *B,* constitute a joint stock of 300 dollars, whereof *A* had 200 dollars and *B* 100. They gain 150 dollars in trade. What is each person's share of the gain? (p. 87)

Through model examples, the Sterrys showed how such tasks could be calculated with decimal fractions. They made a determined effort to link the new decimal currency to the formal study of decimal fractions—but, such an approach was unusual, and perhaps that was why Sterry and Sterry's (1795) book was not published beyond the first edition (Karpinski, 1980).

A Textbook Compiled by Sundry Teachers in and Near Philadelphia (1800)

According to Louis Karpinski (1980), a textbook prepared by John Todd, Zachariah Jess, William Waring and Jeremiah Paul with the assistance of "sundry teachers in and near Philadelphia" (p. 97) was published in 1791. Although, there are no extant copies of the 1791 edition, there are extant copies of editions which were published in 1794, 1796, 1797, 1799, as well as of 15 other editions which appeared well into the nineteenth century. What made the book interesting, historically, was the claim, on the title page, that it was a "practical arithmetic" prepared by practicing teachers.

The earliest edition that we were able to locate was published in 1800, and of particular interest were the sections on vulgar fractions, decimal fractions, and Federal money. Analysis of these sections revealed that aside from a summary of the relative values of the new Federal coins, and some additions, subtractions, multiplications and divisions of Federal money, the book was essentially no different from the kind of school arithmetic found in standard British arithmetics. Almost all of the practical money problems were expressed in sterling pounds, shillings, pence and farthings. Simple and compound interest tasks involved sums of money expressed in sterling rather than Federal money. That remained true of later editions of this book.

The chapters on vulgar fractions appeared after chapters on equation of payments, barter, loss and gain, and fellowship—and the chapters on decimal fractions appeared after the chapters on vulgar fractions. In other words, it was likely that in schools most students who used this book never got to study vulgar or decimal fractions formally. All students who used the book would have been introduced to Federal arithmetic, but only briefly. Almost all word problems involving money calculations were expressed in terms of pounds, shillings, pence and farthings.

Erastus Root's (1795) Textbook

The title of Erastus Root's (1795) text, *An Introduction to Arithmetic for the Use of Common Schools*, would have been attractive to some teachers, as would the following statement in Root's preface:

> To be candid, fellow citizens, the object of this publication is to furnish common schools, with an easy, accurate and cheap volume, containing all the arithmetical knowledge necessary for the farmer or the mechanic. The *manner* may be new if the *matter* is not. Several very excellent treatises on arithmetic have lately been published; yet none of them seem to be exactly calculated for common schools. The size and consequent dearness of some, forbids their general use; while the deficiency and unnecessary learning of others, ought to exclude them. Transatlantic authors will no longer do for independent America. We have coins and denominations of money peculiar to ourselves—In these our youth ought to be instructed and familiarized. The simplicity alone, of this our Federal money, is its sufficient recommendation. Its denominations are the simplest possible—being purely decimal. Almost two centuries have elapsed since the invention of decimal arithmetic; yet never, till lately, has it been applied to the weights, measures, or monies of any nation. But it remained for the United States to make the beginning. (pp. v–vi)

The message seemed to be—the other arithmetics are too expensive, and too difficult, and this arithmetic will give you the kind of arithmetic which you, as an American, will need to know. Root's first edition had only 105 small-sized pages, and early in the book there were nine successive pages totally devoted to "Federal Money" (pp. 20–28). Later pages also focused on the arithmetic of Federal money (e.g., page 49 and page 64, were both concerned with reduction). The writing was clear, and perhaps that is why the book went to 10 editions, the last being issued in 1814. And, as he indicated in his preface, Root was prepared to compromise on the currency question: "I have given many of the examples in pounds, shillings and pence—supposing it necessary to instruct our youth in the *old way* [original emphasis] for some time yet to come. The customs of a great nation cannot be wholly changed in a month, nor a year" (p. vi). So far as both coinage and weights and measures were concerned, this statement was prophetic.

But, as an introduction to mathematics, Root's little book was deficient. He totally omitted any discussion of vulgar fractions because they were "not absolutely necessary" (p. vi). What Root did not seem to recognize was that from a mathematical perspective, fractions are not only an important component of the number system, but they can also be usefully applied in many daily situations. He was certainly not the last person to maintain that fractions should not be part of the common-school arithmetic curriculum, but all attempts to rid schools of the bogey of fractions have failed because, in fact, they *are* mathematically important and because they *can be* useful. Furthermore, any parent in the 1790s who wanted a child to proceed to college, or to be easily able to get a job, would have been concerned if they found that the arithmetic curriculum at the school which their child attended did not include fractions.

Peter Tharp's (1798) Textbook

This small-sized 120-page book was not published beyond the first edition. Tharp designated himself, on the title page, as "Math," which presumably was meant to imply that he taught students mathematics on a private basis. He lived in the town of Marlborough, in the State of New York. In the preface to his book he commented that he had "laid down a very plain and concise rule for reducing the currency of each state into Federal money, and the contrary" (p. iii).

Tharp was another writer who chose not to include a section on vulgar fractions. However, he gave a strong place to both decimal fractions and to Federal currency. Understandably, though, because Congress had decided against introducing a Federal system of weights and measures, all of the compound operations tasks that he included were based on traditional units.

From an education perspective, Tharp's presentation of material often left much to be desired. Thus, for example, the method he used to solve the problem: "What is the interest of 27.5 dol. for five months at 7%?" was not described verbally—rather, it was shown as follows:

$$
\begin{array}{r}
27.5 \\
\underline{.07} \\
1.925 \\
\underline{.4166} \\
11553 \\
11553 \\
1952 \\
\underline{7700} \\
.8019553 \text{ dol.} \\
\underline{10000} \text{ (Tharp, 1798, p. 59)}
\end{array}
$$

Tharp (1798) expected his readers to work out that the .07 on the second line corresponded to the 7% interest, and that the .4166 on the fourth line corresponded to the five months (or 5/12 of a year); the long-multiplication of 1.925 by .4166 should have shown 1925 and not 1952 on the fourth-bottom line; and it was not clear what the 10000, on the last line, meant.

From this example, it is easy to understand why Tharp's book was restricted to one and only one edition.

Chauncey Lee's Plea for Action on Decimalization of Weights and Measures

Who was Chauncey Lee?

An author who offered a particularly radical outlook on the arithmetic curriculum was Chauncey Lee, and his case will now be looked at in some detail. Chauncey Lee was born in Connecticut in 1763, and died in the same state in 1842. He graduated from Yale College, and after practicing Law for some years turned to Theology and was licensed to preach in 1789. He then turned to Teaching and, in 1796 became Principal of Lansingburgh Academy, a new school in Lansingburgh (now part of Troy, New York). In 1797 a 300-page arithmetic textbook (*The American Accomptant being a Plain, Practical and Systematic Compendium of Federal Arithmetic; in Three Parts; Designed for the Use of Schools, and Specially Calculated for the Commercial Meridian of the United States of America*) written by Lee was published (Sprague & Lathrop, 1857). Its publication had been financed by a long list of subscribers, but apparently the venture was not successful because only one edition of the book was ever published (Karpinski, 1980). The book has been well remembered by historians, because it has been claimed that in it Chauncey Lee was the first author to use the now familiar dollar sign ($). But that claim has been challenged (see Fanning & Newman, 2011).

Lee's (1797) *The American Accomptant*

If one reads Lee's lengthy (38-page) introduction to *The American Accomptant* one can get a good idea of why the publication did not go beyond its first edition. The longer version of the title of the book inferred that its author was concerned to write an arithmetic suited to the needs of North American school students, and in his introduction he made it clear that he intended his textbook to live up to its title. He did not name Nicolas Pike in the introduction, but it was obvious from what he wrote that he felt that existing arithmetic texts were inappropriate for school children. From a historical perspective, Lee's introduction is, perhaps, the most important extant statement on the weaknesses inherent in standard approaches to eighteenth-century arithmetic education.

Lee was prepared to tread where previous authors of arithmetic texts had rarely trod. Thus, for example, he recommended the use of the decomposition method of subtraction, rather than "equal additions" (Ellerton & Clements, 2012). Most importantly, for the context of this book, he recommended that a whole new decimalized system of weights and measures be introduced, and he provided details for a possible system that might be used. He took a stand against the use of vulgar fractions—on page *ix,* he argued that "these absurd, untoward fractional numbers" needed to be "banished from practice and the several denominations in all commercial tables of mixed quantities conformed to our Federal money, and established upon a decimal scale." He then pointed out that "to accomplish all this is a talk too great for any individual in a republican government" (p. *ix*). What was needed, he said, was "the arm of Congress to effect it," and it was "equally to be hoped and expected, that their wisdom and patriotism will not be inattentive to so important an object of legislation" (p. *xix*). Lee, then put forward his system of units (pp. *xx* to *xxvi*). After pointing out that "an unnecessary multiplication of the tables of compound quantities will not facilitate the study or practice of arithmetic, but have a contrary effect" (pp. *xxvii*), he warmed to his theme:

> And, let me ask, what real necessity can there be of having such a diversity of weights? What even imaginary necessity, abstract from the current of arbitrary custom and habit? What benefit from it to society in general, or to the tuition of schools in particular? What good purposes are answered by it in the transaction of any kind of business, or in the operation of any arithmetical calculation whatever, which would not be as well, and on the whole much better answered, by reducing them all to practice to a single standard; and ascertaining the gravity of gold, iron, medicines, and all kinds of substances, now classed under three different sorts of weights, by one common table of weights, distinguished and dignified by the name of *American weight*? (p. *ix*)

Lee (1797) then bravely prepared the way to introduce his own scheme for a decimalized system of weights and measures. He argued that Congress had shown foresight in putting into place a decimalized form of currency, and what was needed now was a system of weights and measures which was consistent with the new currency.

Lee then proposed that "Federal Avoirdupois" be based on the relationships" "10 drams make 1 ounce; 10 ounces make 1 pound; 100 pounds make 1 hundred weight; and 10 hundreds make one thousand" (p. *xxi*). So far as "Federal Troy Weight" was concerned, he pointed out that by a 1793 Act of Congress the weight of the American dollar was called a "pennyweight," one-tenth of that, a "cent," and one-tenth of a cent, a "mill." He then proposed that 10 cents should be 1 grain, 10 grains 1 pennyweight, 10 pennyweights 1 ounce, and 10 ounces 1 pound (p. *xxii*). He also put forward a "Federal Apothecary Weight," by which 10 grains equaled 1 scruple, 10 scruples would be 1 dram, 10 drams 1 ounce, and 10 ounces, 1 pound" (p. *xxv*), and showed decimalized tables for liquid measure, dry measure, long measure, and cloth measure.

Lee (1797) realized that, ultimately, it was not really his task to be putting forward such a radical proposal. He wrote:

I need not be reminded that it becomes not a private individual, in a great Republic, to dictate rules and reforms of this kind: I am not so weak as to aspire to it; but only to exercise the republican private privilege of *proposing* [original emphasis] what the more enlightened public may judge of, and candor will not reject without reason. (p. *xxix*)

In sections on weights and measures, Lee included tables of traditional measures and also tables of his own proposed measures. This might have been confusing for some students and teachers. He also showed how the arithmetic of decimal fractions could greatly simplify calculations for weights and measures, and even used his dollar sign ($) when doing it (see, for example, Lee, 1797, p. 87, reproduced as Figure 5.2). Notice that in Figure 5.2 the price of a yard of velvet is given as 3 dollars, 55 cents, and 7 mills, which was recorded as $3.55 7.

Although Lee's thinking about decimalization was ahead of his time, it was politically and educationally naïve. Unlike Thomas Jefferson, Lee failed to recognize that the time had passed when Congress would accept legislation creating a decimalized Federal system of weights and measures. Although attempts had been made to bring the matter to a vote at various times between 1790 and 1792, and again in 1795 and 1796, it had never formally been committed to a vote. Indeed, according to Boyd (1961), the debate in 1796 had seen opponents to the idea treating the proposal "with levity" (p. 617). There was no way the nation would accept Lee's proposals for a unified system, and therefore there was little chance that Lee's textbook, which assumed that his proposals would be in use, would be supported by the public.

Jefferson continued to believe strongly in the desirability of a unified system, but recognized that, politically speaking, he had lost his opportunity. In 1801, when he first became President, Jefferson seemed resigned to the likelihood that the political forces that would line up in Congress against any attempt to introduce such a national system would prove to be too strong. The sun of the new era which had dawned in 1775, had reached its high point around 1790, but was now on the way to a quiet sunset. And, so far as mathematics education was concerned, the new day had brought forth a decimalized system of coinage, but not a decimalized system of weights and measures. Jefferson had recognized that fact by the mid-1790s, but apparently Chauncey Lee had not.

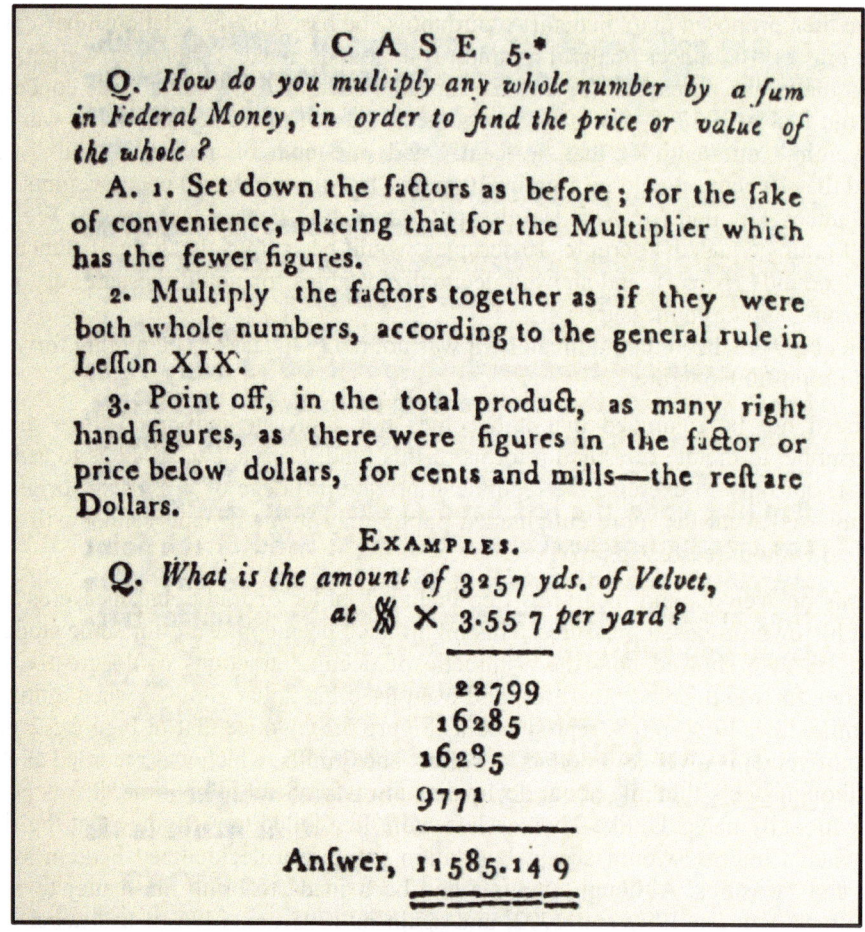

Figure 5.2. Chauncey Lee (1797), multiplying $3.55 7 by 3257 (p. 87).

Influence of Thomas Dilworth's *The Schoolmaster's Assistant* in the 1790s

Of all the arithmetic textbooks used in North American schools during the 1790s, the most popular was that written by Thomas Dilworth, an English cleric. Before he died in 1780 Dilworth had authored numerous textbooks which had sold well in Great Britain. The most popular was his *New Guide to the English Tongue,* but there were also textbooks on arithmetic, bookkeeping and geography (see advertisements in Dilworth, 1784, for details). According to Louis Karpinski (1980), 18 different editions of Dilworth's arithmetic textbook, *The Schoolmaster's Assistant,* were published in the United States of America between 1790 and 1800. Before that, between 1773 and 1790, no less than eight new editions of Dilworth's *Arithmetic* had been published in the United States.

One might think that the popularity of Dilworth's *Arithmetic* in the United States at this time is particularly surprising given that Dilworth's publishers made by no effort to "Americanize" the editions published in the United States. Thus, for example, in a 1797 edition published in New London, Connecticut, Dilworth's 4-page preface is dedicated to the "revered and worthy schoolmasters in Great Britain and Ireland" (Dilworth, 1797, pp. vii–x). There was not one mention of the United States in the preface. In the sections on currency there was no mention of American dollars and cents and, except in the section on Exchange, all money calculations were based on the sterling pounds, shillings, pence and farthings. Perhaps more than anything else, the continued use of Dilworth's *Arithmetic* during the 1790s was due to its popularity in the 1770s and 1780s—the first American edition to be published in North America had appeared in 1773. In the 1790s, and well beyond that, teachers who had taught arithmetic either from Dilworth's *Arithmetic*, or from notes from Dilworth's text which had been copied into cyphering books, found it convenient to continue to keep on teaching the kind of arithmetic that they themselves had learned at school.

Whatever the reason, the fact that Dilworth's *Arithmetic* was more popular than any other arithmetic textbook in North American schools in the 1790s meant that in schools which relied on Dilworth's approach to arithmetic, many students would never study decimal fractions. That was because the section on decimal fractions in Dilworth's *Arithmetic* did not appear until sections on whole-number arithmetic, and on vulgar-fractions arithmetic, had been concluded. Thus, for example, it was not until page 123 in the 1797 edition of Dilworth's *Arithmetic*, published in New London, Connecticut, that decimal fractions were first mentioned. Remarkably, even an 1802 edition of Dilworth's *Arithmetic* which had been especially prepared for the American market by Robert Patterson, Professor of Mathematics at the University of Pennsylvania—someone who occasionally corresponded with Thomas Jefferson on issues associated with decimalization—made no reference to decimal currency. That was still the case even in Part III of the text, which was devoted entirely to decimal fractions (Dilworth, 1802).

One of the consequences of the popularity of Dilworth's *Arithmetic* was that students were asked to learn cumbersome methods for solving problems that could have been solved much more simply if the topics had been dealt with in a different order. Consider, for example, Dilworth's recommended method for finding the interest of 200 pounds for 3 years and ¾ at 5 per cent per annum (see Figure 5.3 from Dilworth (1797, p. 64). Dilworth had not yet dealt with vulgar fractions, or percentage, in his book, and so the sequencing of the curriculum was poor. Immediately before the problem, Dilworth (1797) had elaborated the method for solving Case II problems in the following way:

> Q. How do you find the interest of any sum for ¼, ½, or ¾ of a year, besides the number of years given in the question?
>
> A. For ¼ of the year, take a fourth part of the interest for one year; ½ of the year, take half of the interest for one year; for ¾ of the year, take the parts compounded of ¾ and add them to the interest for the rest of the time; the answer will be the interest required. (p. 64)

How Dilworth applied this method to find the solution to the problem of finding the interest of 200 pounds for 3 years and ¾ at 5 per cent per annum is shown in Figure 5.3. One wonders how any student being introduced to the notion of simple interest could have followed either Dilworth's explanation of the method, or the working shown in his illustrative example.

Federal Currency, Sterling Currency, Weights and Measures, and Vulgar and Decimal Fractions in Implemented Arithmetic Curricula 1792–1800

The above discussion of the different emphases given by North American authors of arithmetic textbooks, during the period 1788–1800, to decimal fractions, vulgar fractions, decimal currency, sterling currency, and weights and measures, raises the question of what was actually being studied in the name of school arithmetic in the schools.

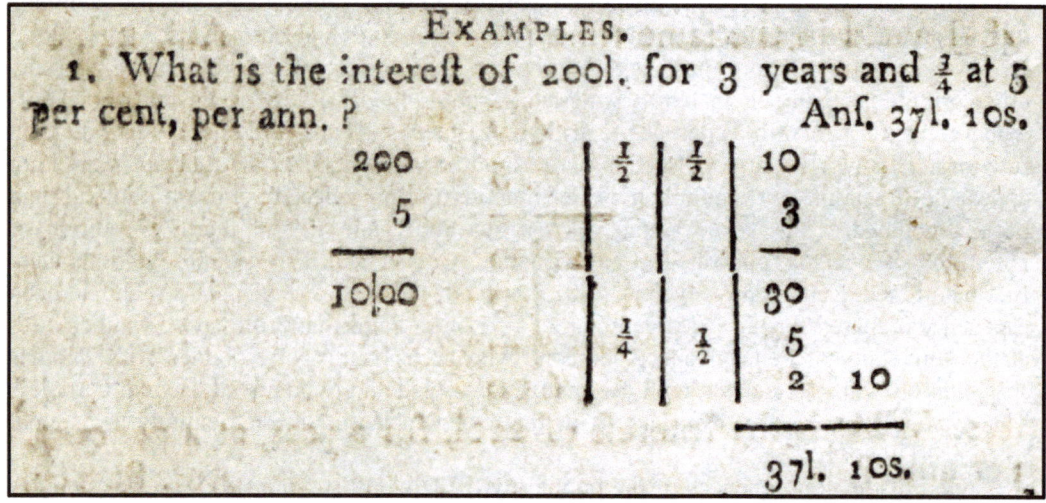

Figure 5.3. Finding "the interest on 200 pounds for 3 years and ¾ at 5% per annum."

In Table 5.1 a summary of the emphases found in 21 cyphering books, from the Ellerton-Clements collection is given, each of which was prepared, at least partly, during the period 1792 through 1800. The 21 cyphering books were prepared in different parts of the United States. The first five entries in Table 5.1 have been separated from the other 17 because students who prepared them did not mention decimal currency on any of their pages.

The 21 cyphering books provide commentary on the implemented school mathematics curriculum during the years immediately after 1792, the year when the United States Mint was established. Three of the columns in Table 5.1 have the same headings as columns in Table 2.2 (in Chapter 2), which covered the period 1742–1791, and comparison of trends might be of historical interest. Entries in Table 5.1 suggest the extent to which decimal currency was studied in schools in the years immediately following its introduction. The entries also reveal the extent to which weights and measures and associated tasks formed part of U.S. school arithmetic curricula just before the end of the eighteenth century.

Table 5.1
Summary of Arithmetical Topics in 21 North American Cyphering Books, 1792–1800

Name and Year(s)	Numeration and Four Operations (%)	Federal (Decimal) Money (%)	Sterling and Other Money (%)	Weights and Measures (%)	Rules of Three (%)	Not Money, Not Weights and Measures (%)	Vulgar Fractions (%)	Decimal Fractions (Pure) (%)
Hough, 1795–1798	14	0	44	9	33	0	0	0
Northboro, MA 1795–1798	0	0	5	20	7	0	54	14
May J. Underhill 1796	5	0	46	24	15	10	0	0
Thomas Williams 1797	14	0	28	35	23	0	0	0
Hugh Ross 1797–1802	0	0	11	0	0	47	42	0
Unknown Student c. 1792	22	6	44	25	0	3	0	0
Olney Brayton 1792–1794	9	2	60	23	0	6	0	0
Elisha Harris 1794–1797	0	36	6	7	43	0	0	9
John Kerling 1795	21	2	23	35	14	5	0	0
Solomon True 1795	17	2	15	19	11	0	0	0
Warren Green 1795–1800	28	19	14	17	23	0	0	0
William Senter 1795–1801	1	1	41	8	37	1	0	11
Benjamin Tolman 1796	10	2	42	24	22	0	0	0
Lucy Starr 1797	10	21	17	36	16	0	0	0
Levi Johnson 1798–1801	27	3	24	17	29	0	0	0
Jonathan Valent- tine, 1798–1802	18	4	16	42	10	0	0	10
Warner Green 1799-1800	4	16	38	11	22	9	0	0
J. D. Hatfield c. 1800	0	13	60	17	0	10	0	0
Thomas Cooper c. 1800	0	13	4	8	29	0	44	21
Unknown Author c. 1800	0	20	3	2	7	47	21	0
Author Unknown 1800–1805	1	11	27	18	12	22	8	0
Mean %	10	8	27	19	17	8	8	3

The entries in Table 5.1 begin to throw light on likely answers to the following questions:

1. Did some students study decimal currency but not decimal fractions?
2. Did some students study decimal fractions, but not decimal currency?
3. Were vulgar fractions more commonly studied than decimal fractions?

4. To what extent after 1791 did implemented arithmetic curricula in U.S. schools continue to emphasize sterling currency?
5. Were implemented curricula similar to the author-intended curricula revealed by the above analyses of texts by U.S. authors Nicolas Pike, Benjamin Workman, Consider and John Sterry, Erastus Root, Peter Tharp, and Chauncey Lee?

Did Some Students Study Decimal Currency but not Decimal Fractions?

The answer to the first of these five questions is clearly "Yes"—as many as 12 of the 21 students who prepared the cyphering books which are represented in Table 5.1 were in that category. The students were living in a nation in which a Federal decimalized currency had only recently been introduced. Many teachers had never learned, or taught, decimal fractions as a topic. Rather than struggle to learn the concepts associated with decimal fractions, some teachers could have deemed it to be easier merely to learn to add, subtract multiply, and divide Federal money—for that was like doing "similar" operations on natural numbers.

The 202-page cyphering book prepared by Warner Green, of Rhode Island, was one of the manuscripts summarized in Table 5.1, and Figure 5.4 reproduces a page showing Warner's handwritten solutions to the following five "division" word problems:

- 189 dollars 46 cents divided by 46.
- If 372 yards of cloth cost 370 dollars, what will one yard of ditto cost?
- Gave 34 dollars 25 cents for 3 cwt of sugar; what is that per lb?
- Bought 9½ gallons of brandy for 19 dollars, 50 cents, what is it per gallon?
- If 10 cwt of beef cost 4 dollars, how much is that per cwt?

Immediately before these problems, Warner had written the following rule:

Divide the price of the quantity as in which numbers: if the price be dollars only and you have brought them all down in the dividend place a comma in the quotient, and add two cyphers to the remainder; and divide as before and the figures on the right hand of the comma in the quotient will be cents; if there be still a remainder place another comma in the quotient, add a cypher to the remainder, and divide again and the figure on the right hand of this last comma will be mills.

This quotation provides an example of part of an aspect of what has been called the IRCEE ("Introduction-Rule-Case-Example-Exercise") genre (Ellerton & Clements, 2012), which so often characterized statements in cyphering books. There can be no guarantee that Warner understood the rule (which he probably copied, or wrote down when the words were dictated to him). Nowhere in Warner's large cyphering book was there an explanation of why the rule "worked." Writing "rules without reason" (Skemp, 1976) was often the approach adopted during the cyphering era in North America.

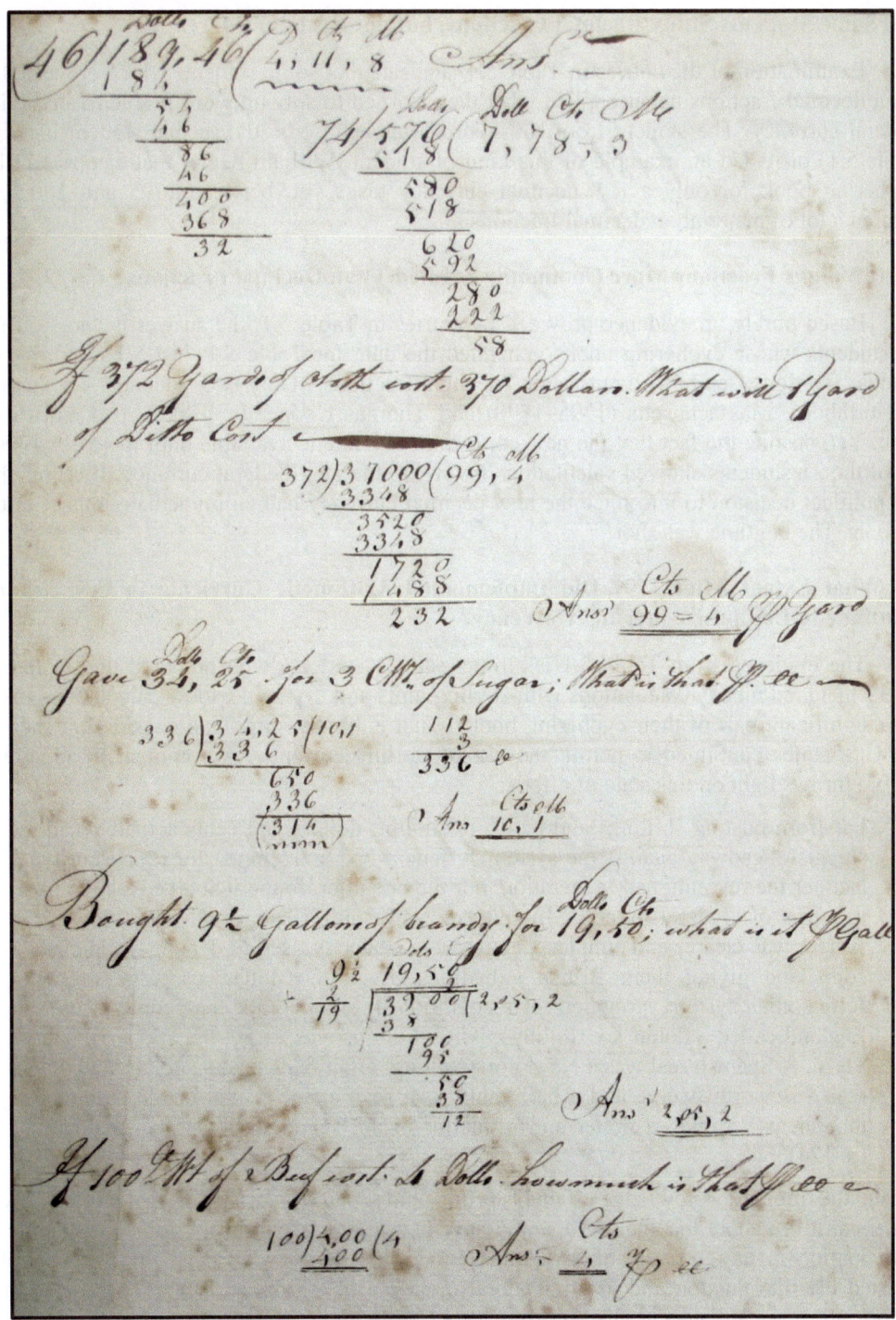

Figure 5.4. Warner Greene's solutions to division word problems with Federal money.

Did Some Students Study Decimal Fractions, but not Decimal Currency?

Examination of the entries in Table 5.1 suggests that some students who were asked to learn decimal fractions never applied what they learned to solving word problems involving Federal currency. The student from Northboro, Massachusetts (the second student listed in Table 5.1) provided an example of this kind of student. William Senter made entries in his cyphering book for only a few decimal currency tasks yet, between 1795 and 1802, he entered a lot of material on decimal fractions.

Were Vulgar Fractions More Commonly Studied Than Decimal Fractions?

Based purely on evidence provided by entries in Table 5.1, the answer is "no." Of the 21 students whose cyphering books generated the data for Table 5.1, just 5 had entries on "vulgar fractions," and 5 had entries on "decimal fractions." Only two students—the student at Northboro, Massachusetts (1795–1798) and Thomas Cooper (c. 1800)—had entries on both. Yet, despite the fact that the new coins did not become available until the early 1790s, 16 of the 21 students showed calculations involving the new Federal currency. It seems that the political decision to introduce the new decimal currency had an immediate impact in the schools. The lag time was short.

To What Extent After 1791 Did Implemented Arithmetic Curricula in U.S. Schools Continue to Emphasize Sterling Currency?

The evidence from Table 5.1 is strong—all 21 students who prepared the cyphering books included money calculations with sterling units and, for some, these calculations made up a significant part of their cyphering books. That is hardly surprising considering each of the U.S. states continued to permit the use of sterling currency. A comment by Goodwin (2003) throws light on this state of affairs:

> Far from ousting shillings and pence overnight, dollars and cents actually made sluggish headway against the system Jefferson had condemned for its obscurity. Neither the revolutionary generation, nor the one after, found it at all easy to think in terms of "federal money." In spite of Jefferson's argument that the decimal system was clearer and simpler, people were perfectly used to dealing in Spanish coins and giving them British valuations. Even the dollar, capstone of the Jeffersonian system, struggled for recognition. Into the 1850s the people of New England called a dollar six shillings. Nine shillings was $1.50; ten and six meant $1.75. A Spanish real was a New York shilling; eight reals made a dollar, and one real was worth twelve and a half cents. Ten reals made $1.25, or ten shillings, though in Virginia, a dollar and a quarter meant seven shillings and sixpence. (p. 124)

Given the intimate knowledge of the sterling and Spanish currencies that almost every American from about the age of 10 would have developed as a result of school arithmetic or participation in day-to-day monetary transactions, it was not surprising that "the people" resisted the introduction and use of Federal money. But entries in Table 5.1 indicate that, despite all, Jefferson's decimalized currency was being given its due attention in the schools.

Were Implemented Curricula Similar to the Author-Intended Curricula Implied in the Texts by Nicolas Pike, Benjamin Workman, Consider and John Sterry, Erastus Root, Peter Tharp, and Chauncey Lee?

Although all of the textbook authors included chapters on decimal fractions, only 5 of the 21 cyphering books examined included pages on decimal fractions (as distinct from decimal currency). None of the cyphering books dealt with algebra, yet Nicolas Pike and Consider and John Sterry had included significant sections on algebra in their textbooks. And, no-one followed Chauncey Lee's idiosyncratic treatment of weights and measures. So, the short answer to the question is that the new textbooks did not make much of an impact on the implemented curriculum—however, 16 of the 21 cyphering books did include Federal money calculations, and such calculations appeared in all of the afore-mentioned textbooks.

Weights and Measures in Teacher-Implemented Curricula

Entries in Table 5.1 indicate that implemented curricula usually included a significant emphasis on weights and measures. Not only were there pages devoted to compound arithmetic calculations with respect to length, area, volume, capacity, weight, time, and angles—which accounted for most of the entries in the column headed "Weights and Measures" in Table 5.1—but also entries in the column headed "Rule of Three" often derived from tasks involving weights and measures, and money.

Sections in cyphering books with headings like "Cloth Measure" and "Dry Measure," etc., are made up of simple tasks made difficult for school children by the multitude of numerical relationships between units which needed to be taken into account. Figure 5.5, for example, is taken from the section in Lucy Starr's (1797) cyphering book concerned with compound operations involving addition. Under "Cloth Measure," Lucy listed the relationships 4 nails (Nl) make one quarter, 4 quarters make 1 yard (Yd), and 5 quarters make 1 ell (E). Those between-units numerical relationships were summarized above her first example—on top is 10, 4, 4, and immediately below that is Yd, Qr, and Na. Then, in three different examples, six quantities of cloth were summarized, and it was expected that these would be added. Thus, with the first column (on the right), the sum is 12 (nails), which corresponds to 3 quarters. Then, with these 3 quarters "carried," the second (middle) column sums to 15 (quarters), which is 3 yards and 3 quarters left over—so "put down the 3 and carry the 3." Then, the left column sums to 197 (Yds), which is not the 287 (yards) shown). Then, Lucy showed a check, which apparently assured her that her calculations were right (when in fact they were wrong). In the middle calculation, also involving yards, quarters, and nails, Lucy once again recorded an incorrect answer, but her check suggested to her that she had been correct. The answer to the third task (on the right side), involving ells, quarters, and nails, was also incorrect but checked as correct. One is led to ask, who did the checking?

With the exercises on dry measures, six amounts of dry measure were summarized, and addition was expected to occur. The relationships to be invoked were 2 pints make 1 quart, 8 quarts make 1 Peck, and 4 Pecks make 1 bushel. This time Lucy's calculations were correct.

Lucy included many pages with setting out just like that shown in Figure 5.5. First, she added quantities for all the different kinds of measures, then she subtracted them, then she multiplied quantities by fixed numbers, then she divided quantities by fixed numbers. And, after all that she moved on to reduction of quantities, for which she recorded solutions to tasks like "Bring 2039040 grains into pounds" and "In 4 leagues how many feet?"

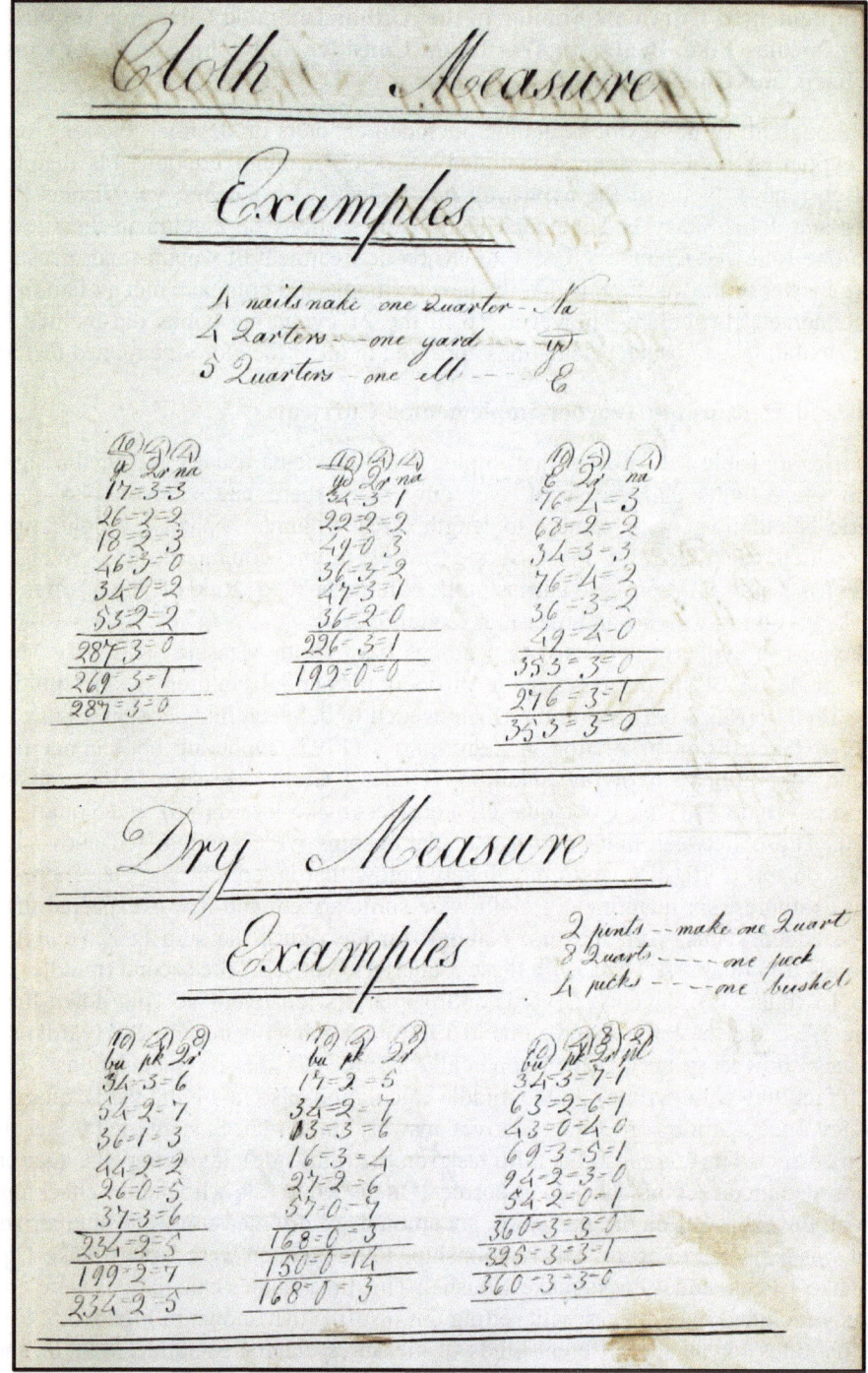

Figure 5.5. Lucy Starr (1797) "adds" cloth measures and dry measures.

From an education perspective, there are two ways of looking at these compound operations tasks. One might see them as educationally worthwhile, because students had to keep their wits about them—constantly reflecting on which units were involved, and how the relevant units were related. From another perspective, however, one might see them as educationally wasteful because they took up so much of the time allocated to the study of arithmetic. Those, like Jefferson, who were seeking to create an integrated, decimalized system of weights and measures, saw the issues from this latter perspective. Those who saw things like this asked: Why not standardize relationships between units, so that they are always based on ratios of powers of ten? Then students would be able to concentrate on the quantities involved rather than on the different relationships between units. Certainly, entries in Table 5.1 suggest that at the very time Jefferson was arguing for a new coordinated system of weights and measures, much time was being spent on weights and measures in the schools.

Vulgar Fractions, Decimal Fractions, and Federal Money in Arithmetic Textbooks 1801–1810

The new century heralded the publication of many relatively inexpensive arithmetic textbooks authored by U.S. citizens. Some of these would go through many editions, and would be more suited than Nicolas Pike's textbooks to the needs and abilities of "ordinary" school children. The most popular arithmetic textbooks written by U.S. citizens in the first decade of the nineteenth century were authored by Zachariah Jess (first edition 1799, and six editions by 1810), Nathan Daboll (first edition 1800, and 10 editions by 1810), Michael Walsh (first edition 1801, and eight editions by 1810), and Daniel Adams (first edition 1801, and six editions by 1810) (Ellerton & Clements, 2012; Karpinski, 1980).

Thomas Dilworth's *The Schoolmaster's Assistant* remained the most widely used arithmetic textbook, however, and 20 separate American editions of it were published between 1800 and 1810. Nicolas Pike's two arithmetics had new editions between 1801 and 1810—two for Pike's original, larger arithmetic, and five for his *Abridgement* (Karpinski, 1980).

Adams, Daboll, Walsh and Jess left no doubt that their author-intended arithmetic curriculum included a much greater emphasis on decimal currency than had been given in the textbooks available in the 1790s. Adams (1801) emphasized the importance he attached to Federal money when, on his title page, he included "FEDERAL MONEY" as one of the seven major themes of his book. In his chapter on "Fractions," Adams devoted just two pages to vulgar fractions (pp. 75–76) but 11 pages to decimal fractions (pp. 76–86). He then devoted an additional 11 pages (pp. 87-97) to Federal money.

Nathan Daboll (1800) included a three-page specific section on vulgar fractions (pp. 74–77) and a twelve-page section on decimals fractions which incorporated aspects of the new Federal currency (pp. 77–88). In his preface, Daboll included the following historically important statement relating to curriculum development:

> In the arrangement of fractions, I have taken an entirely new method, the advantages and facility of which will sufficiently apologize for its not being according to systems. As decimal fractions may be learned much easier than vulgar fractions, and are more simple, useful, and necessary and soonest wanted in more useful branches of arithmetic, they ought to be learned first, and vulgar

fractions omitted until further progress in the science shall make them necessary. It may be well to obtain a general idea of them, and to attend to two or three easy problems therein; after which the scholar may learn decimals, which will be necessary in the reduction of currencies, computing interest, and many other branches.

Besides, to obtain a thorough knowledge of vulgar fractions is generally a task too hard for young scholars who have made no further progress in arithmetic than reduction, and often discourages them. (pp. v–vi)

This echoed what Erastus Root (1795) had written in the preface to his textbook, *An Introduction to Arithmetic for the Use of Common Schools.*

So far as vulgar fractions, decimal fractions, and Federal currency were concerned, Michael Walsh (1801) adopted a different strategy. He introduced "Federal money" early in his book (on p. 19), and then immediately introduced sterling currency ("English money") as well. Then, whenever a context involved money, he offered methods for both forms of money. For example, Figure 5.6, reproduced from page 106 of Walsh (1801), shows the last of six pages on simple interest with sterling currency" and the first of 12 pages on simple interest in "Federal money."

Walsh had close contacts with New England merchants, and his dual approach was designed to win the support of merchants. In fact, in the first decade of the nineteenth century, many money transactions involved sterling currency and many involved Federal money. From the title of Walsh's (1801) textbook—*A New System of Mercantile Arithmetic: Adapted to the Commerce of the United States, in its Domestic and Foreign Relations; with Forms of Accounts, and Other Writings Usually Occurring in Trade*—it could not been clearer what he was trying to do. At the front of his book there were recommendations from merchants in Newburyport, Boston, and Salem, which were all major sea ports engaged in trade with U.S. markets, Europe, and the East and West Indies. Young men who were destined for employment by traders in towns like Newburyport, Boston, and Salem often prepared cyphering books that were based on Walsh's textbook.

The preceding comments, on popular arithmetic textbooks used in the United States of America during the first decade of the nineteenth century, begged the question—what proportion of school students studying arithmetic actually owned and used an arithmetic textbook. Elsewhere (Ellerton & Clements, 2012), we have provided evidence that most students neither owned a textbook nor were in a position to refer to one. Often, students were able to consult another ("parent") cyphering book—owned, perhaps, by the teacher or by one of the student's parents or someone else in the family. Sometimes it seemed that notes in cyphering books were copied from textbooks—especially when the wording of rules, or examples, could be found in the textbooks. But, even so, one should not assume that the student had access to that textbook.

In earlier chapters of this book we have shown that textbook-intended arithmetic curricula often did not match implemented curricula—as indicated in cyphering books. It follows that one cannot be confident about what the implemented arithmetic curriculum actually was for a particular teacher and for his or her students, until an analysis of relevant cyphering-book data has been carried out.

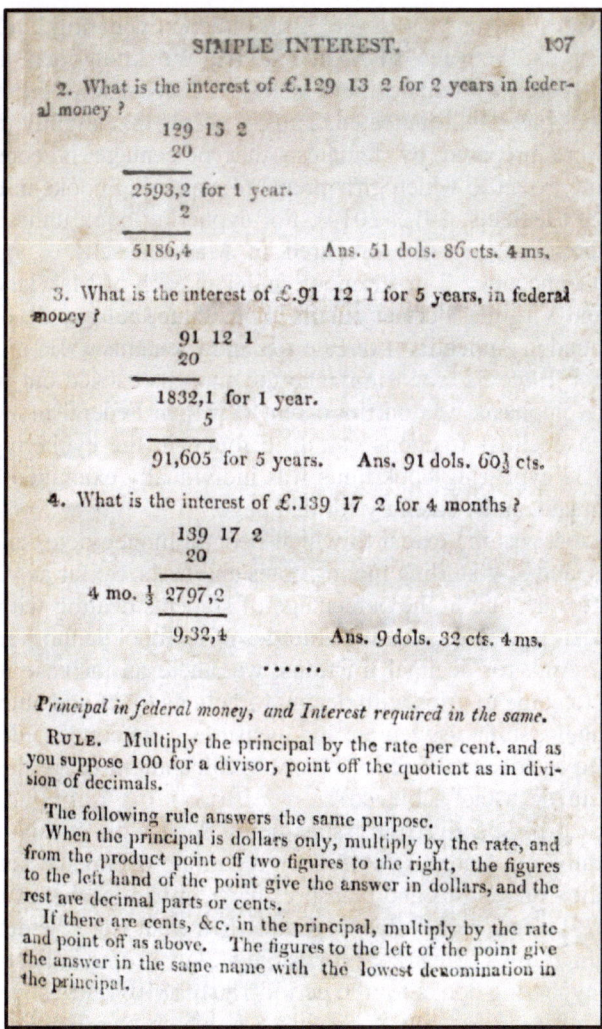

Figure 5.6. A page from Michael Walsh's (1801, p. 106) arithmetic textbook displaying sterling and Federal currency methods for calculating simple interest.

Cyphering-Book Data on Implemented Arithmetic Curricula 1801–1810

Within the Ellerton-Clements collection of 370 U.S. cyphering-book units are 32 cyphering-book units that were concerned primarily with arithmetic and were prepared in different parts of North America during the period 1801–1810. The term "cyphering-book unit" was used to indicate that each unit was, as far as we know, prepared independently of each other unit. Thus, for example, if the same person prepared two or more cyphering books, and those cyphering books are part of the Ellerton-Clements collection, then they have been counted as just one unit. This distinction, between "cyphering book" and "cyphering-book unit" is made to improve the likelihood that the analysis will yield results that can be generalized across the United States of America for the period 1801–1810.

Table 5.2 provides an analysis of the 32 arithmetic cyphering-book units which were prepared, at least partly, between 1801 and 1810. We know that at least 5 of these manuscripts were prepared by females, and at least 22 by males—the names of 7 persons who prepared the books were not indicated. Thus, it seems that slightly less than 20% of the cyphering books were prepared by females—that percentage is consistent with recent investigations into the extent to which girls prepared cyphering books in the United States of America (Ellerton & Clements, 2012, 2014). For cyphering-book units for which we know where they were prepared, 6 were prepared in Massachusetts, 6 in New York, 7 in Pennsylvania, 3 in Connecticut, 2 in Vermont, and 1 in each of Maryland, New Hampshire, New Jersey, Ohio and Virginia. Further details about some, but not all, of the 32 cyphering-book units can be found in Appendix *A* of Ellerton and Clements (2012).

The analysis for Table 5.2 was similar to the analysis carried out when Table 5.1 was being prepared—the emphasis was on the extent to which Federal money, sterling money, vulgar fractions, and decimal fractions could be found in the cyphering-book units. Each page of each of the 32 cyphering-book units was individually examined by the two authors, and decisions were made on the category for that page.

Of special interest was the extent to which Federal money notes and tasks appeared in the cyphering books, and also sterling money notes and tasks. So far as vulgar fractions were concerned, was there a decline in the percentage of students dealing with vulgar fractions— which might have been expected given the attitudes of textbook authors such as Erastus Root and Nathan Daboll? And, for decimal fractions, was there an increase in the percentage of students attempting to come to grips with decimal fractions (over and above what they did on Federal money)? Finally, so far as weights and measures were concerned, was there evidence that students began to use decimal fractions more in calculations (even though decimalization of weights and measures was not achieved)?

From a historical perspective, the issues being dealt with in this present analysis are important—specifically, did Congress's decision regarding decimalization of Federal money and non-decimalization of weights and measures have any noticeable effect on implemented arithmetic curricula? If a noticeable effect is found, then what were the details of that effect?

Entries in the final row ("Mean %") for Table 5.2 give some indication of the emphases to be found in U.S. cyphering books for the period 1801–1810. Entries in that row and entries in the corresponding "Mean %" row for Table 5.1, which was for the period 1792–1800, have been shown in Table 5.3 (which immediately follows Table 5.2).

Three observations on changes in the implemented curricula for school arithmetic around 1800 are offered. First, during the periods 1792–1800 and 1801–1810—there was a steady increase in the amount of time dedicated, in school arithmetic, to the study and use of Federal money. Before 1792 the mean percentage of space dedicated to entries in cyphering books was, obviously, 0%; then during 1792–1800 it became 9%; and during 1801–1810 it became 18%. This growth seemed to be at the expense of time spent on the various forms of the "Rule of Three." It should be noted that 26 of the 32 cyphering books summarized in Table 5.2 included pages dedicated to Federal money.

Table 5.2
Summary of Arithmetical Topics in 32 North American Cyphering Books, 1801–1810

Name and Year(s)	Numeration and Four Operations (%)	Federal (Decimal) Money (%)	Sterling and Other Money (%)	Weights and Measures (%)	Rules of Three (%)	Not Money, Not Weights and Measures (%)	Vulgar Fractions (%)	Decimal Fractions (Pure) (%)
R. M. Bartlett 1801	0	64	21	4	0	11	0	0
Sarah Peckham 1801	0	35	27	0	0	19	3	17
Noah Burnham 1802-1803	24	0	29	50	0	0	0	0
Isaac Clay 1803	25	0	30	45	0	0	0	0
John Steele 1803–1804	5	0	58	5	33	0	0	0
Stewart Rathbun c. 1805	3	11	22	14	29	0	14	8
Joshua Jacquith c.1805–1810	26	20	17	35	3	0	0	0
Unknown Author c. 1805	0	19	13	19	9	0	0	40
Jacob Richards 1805–1810	5	8	33	33	4	1	7	9
Sarah Mead c. 1806	4	5	14	10	8	17	29	12
Peter Hammond 1806–1807	0	0	82	18	0	0	0	0
Hiram Emery 1807	2	19	38	41	0	0	0	0
Oliver Parry 1807	0	0	34	0	27	1	20	18
Adam Rightnauer 1808	41	13	23	23	0	0	0	0
Henry Kimball 1808–1809	11	20	27	16	2	15	1	7
Isaac Heaton 1809	13	20	13	19	4	15	8	9
Unknown Author 1809	17	27	12	37	7	0	0	0
Barnett Flattery 1809–1810	0	24	31	13	24	0	0	10
Josiah Gardner 1809–1810	14	15	9	33	0	0	19	10
Unknown Female 1810	32	9	15	31	13	0	0	0
Israel Wilson 1810	4	36	11	19	4	8	6	13
Silas Mead 1810–1814	5	0	71	17	3	4	0	0
Jessamine Foshay 1810–1818	12	19	39	28	2	0	0	0
David Calderwood c. 1810	0	43	23	4	20	11	0	0
Jack C. Grover c. 1810	27	32	14	18	9	0	0	0
Sa Lister c. 1810	24	14	24	38	0	0	0	0

(continued)

Table 5.2 (continued)
Summary of Arithmetical Topics in 32 North American Cyphering Books, 1801–1810

Name and Year(s)	Numeration and Four Operations (%)	Federal (Decimal) Money (%)	Sterling and Other Money (%)	Weights and Measures (%)	Rules of Three (%)	Not Money, Not Weights and Measures (%)	Vulgar Fractions (%)	Decimal Fractions (Pure) (%)
John & Charles Scott 1810–1815	0	12	38	3	6	38	2	1
Susannah Lovejoy c. 1810	44	11	15	30	0	0	0	0
Unknown Author c. 1810	7	14	34	29	13	0	0	3
Unknown Author c. 1810	6	16	25	15	8	20	6	4
Unknown Author c. 1810	0	38	12	0	12	0	38	0
Unknown Author c. 1810	0	9	28	22	0	41	0	0
Mean %	11	17	27	21	8	6	5	5

Table 5.3
Mean % of Arithmetical Topics in U.S. Cyphering Books, 1792–1800 and 1801–1810

Period	Numeration and Four Operations (%)	Federal (Decimal) Money (%)	Sterling and Other Money (%)	Weights and Measures (%)	Rules of Three (%)	Not Money, Not Weights and Measures (%)	Vulgar Fractions (%)	Decimal Fractions (Pure) (%)
1792–1800 (from Table 5.1) (Based on 21 manuscripts)	9	9	27	18	17	8	8	3
1801–1810 (from Table 5.2) (Based on 32 manuscripts)	11	17	27	21	8	6	5	5

The second observation is that the proportion of pages dedicated to sterling currency—pounds, shillings, pence and farthings—in cyphering books remained constant, despite the extra attention being given to Federal money. Although that might be deemed to be surprising, the finding testifies to the determination of the states to continue to use their own forms of sterling currency as "legal currency" (Allen, 2009). It might also testify to the unwillingness of many teachers to teach new material. Throughout the period 1792–1810 Thomas Dilworth's *Arithmetic* remained popular—although its popularity was significantly less around 1810 than it had been around 1792—and, of course, in any Dilworth *Arithmetic* there was never any reference to Federal money. All 53 students who prepared the cyphering books summarized in Tables 5.1 and 5.2 included material featuring sterling currency.

The third observation is that, throughout period 1792–1810, only a minority of those students who prepared cyphering books included sections on either vulgar fractions or decimal fractions. Between 1801 and 1810, only 14 of the 32 students who prepared the cyphering books summarized in Table 5.2 included sections on decimal fractions and, most of those who did seemed to be concerned with nothing more than adding, subtracting, multiplying, and dividing decimals. Using decimals when working on non-trivial problem-solving tasks was not common. Nevertheless, during the period 1801–1810 less attention was given to vulgar fractions and more to decimal fractions than was the case during 1792–1800.

Analysis of the cyphering books suggested that there was a belief that Federal money should be presented as a topic in its own right. More general aspects of decimal fractions were only rarely discussed in the cyphering-book entries. That said, in Table 2.2 (in Chapter 2)—which presented an analysis of 21 cyphering books prepared in North-American schools between 1742 and 1791—only 2% of the contents in the cyphering books dealt with vulgar fractions and 2% dealt with decimal fractions. From that perspective, entries in the last two columns of Table 5.3 suggest that between 1792 and 1810 there was an educationally significant leap in the attention being given to vulgar and decimal fractions in the schools, and that was especially the case for decimal fractions.

Indeed, entries in Table 2.2 indicate that only 5 of the 21 students who prepared cyphering books for the period 1742–1791 made entries on general aspects of decimal fractions; in Table 5.1 it can seen that for the period 1792–1800 that proportion remained steady, but from Table 5.2 one can see that for 1801–1810 the proportion increased to 14 out of 32. The increase might have been small, but it happened, and from a lag-time perspective one might argue that Jefferson's success in getting a decimal form of currency accepted as Federal money was enough to make decimal fractions a mainstream aspect of school arithmetic. That issue will be further considered in Chapter 6.

Philosophical, Economic, Political, and Social Influences on Curricula

Lewis White Beck (1966) included Thomas Jefferson among a list of significant "enlightenment" philosophers of the eighteenth century, and although Jefferson's list of philosophical writings is not a long one, the 1776 Declaration of Independence was sufficiently incisive and challenging that few that would question his massive influence on the future of North America. In this book we have seen the mind of this intellectual giant at work seeking to introduce a coordinated and radical scheme that would have linked currency and weights and measures through an integrated decimalized system the likes of which the world had never known. Jefferson sensed that if his scheme were to have a chance of being accepted in a Congress comprising mainly investors and businessmen with vested interests in maintaining, as far as might be possible, the status quo so far as measurement was concerned, he would have to move quickly. But, in 1784 he was posted to Paris, and as a result he "lost" five vital years. Before he left for France he had done much of the preliminary work necessary to get the world's first fully decimalized currency in place, and while he was away the final negotiations for the establishment of a Federal mint were concluded.

On Jefferson's return, President George Washington assigned to him the task of developing and shepherding through Congress a national system of weights and measures. Before he departed for Paris Jefferson had worked out a radical proposal on weights and measures which was consistent with his Federal currency legislation (Boyd, 1961), but in 1790 he sensed that his pre-Paris scheme for an integrated, decimalized scheme might not be approved by Congress. Nevertheless, with the encouragement of Washington, and as a result of long discussions he had had with French luminaries and with local mathematicians and scientists, he submitted a report on weights and measures that was consistent with the new decimalized currency. However, because he feared that it might be the opinion of the representatives that the difficulty of changing the established habits of a whole nation presented "an insuperable bar to this improvement" he submitted an alternative, more conservative plan, "that the house may at their will adopt either the one or the other exclusively; or the one for the present, and the other for a future time" (Thomas Jefferson, "Second State of the Report on Weights and Measures. April 20–May, 1790," in Boyd, 1961, pp. 628–643).

Between April and July, 1790 Jefferson submitted to Congress his three reports on the subject of weights and measures (Boyd, 1961). Some might have thought that the new day which dawned on July 4, 1776, was still young, but Jefferson was very conscious that the sun was setting on his weights and measures proposal. Early in 1790 he seemed to make sense that it was now or never, for at some later date it would be even more difficult to get Congress to agree to a similar scheme. As it happened, neither of the schemes he put forward to Congress was ever voted upon, and the new nation was left with the uncoordinated strands for weights and measures that it had inherited from Great Britain. The fate of one of the most remarkable schemes ever conceived for the coordination of money and weights and measures had been decided by a Congress in which "big business" was in a majority.

Although, Jefferson's system of decimal currency had been approved as "Federal money," the merchants were still entitled to trade within and between states using the previous state currencies involving pounds, shillings and pence. It was assumed that in time the state currencies would disappear to be totally replaced by the new Federal money. The old currencies were collectively termed "legal currency" (McCusker, 1992, 2001) and that expression was often to found in calligraphic headings in students' cyphering books. The terminology of sterling currency continued to be used in the United States for most of the nineteenth century (see, e.g., Robinson, 1870).

No-one realized more than Thomas Jefferson that the introduction of Federal money would have important education effects. Indeed, one of his main arguments in proposing the decimal aspect of the new scheme was that it would empower ordinary people because they would be able to do money calculations much more easily under the new scheme. Jefferson always recognized that there was a strong educational component to his coinage and weights and measures proposals. He was convinced that if decimalized schemes were to be put in place then it would be easier for everyone to learn the kind of arithmetic they would need to be able to identify and master crucially important calculations associated with buying, selling, and measuring.

The analysis of cyphering-book data in this chapter has suggested that Jefferson was right when he argued that the introduction of a decimal form of Federal money would generate important changes in school arithmetic. All of a sudden, a majority of those who studied arithmetic in the schools were doing money calculations using decimal approaches. But, it appears that the students could learn to do Federal money calculations without needing to achieve a relational understanding of decimal concepts (Skemp, 1976). Teachers could ask their students to make calculations with decimal representations of money—involving eagles, dollars, dimes, cents and mills—without generalizing what they were doing to the broader idea of representing numbers using decimal notations. And, meanwhile, the old forms of sterling currency remained alive and well in school arithmetic curricula, as did the old forms of weights and measures.

References

Adams, D. (1801). *The scholar's arithmetic—Or Federal accountant.* Leominster, MA: Adams and Wilder.

Albree, J. (2002). Nicolas Pike's Arithmetic (1788) as the American *Liber Abbaci.* In D. J. Curtin, D. E. Kullman, & D. E. Otero (Eds.), *Proceedings of the Ninth Midwest History of Mathematics Conference* (pp. 53–71). Miami, FL: Miami University.

Allen, L. (2009). *The encyclopedia of money* (2nd ed.). Santa Barbara, CA: ABC-CLIO.

Beck, L. W. (1966). Introduction. In L. W. Beck (Ed.), *Eighteenth-century philosophy* (pp. 1–11). New York, NY: The Free Press.

Bowen, C. D. (2007). *Miracle at Philadelphia: The story of the constitutional convention, May to September 1787.* New York, NY: Little Brown & Co.

Boyd, J. (1961). Report on weights and measures: Editorial note. In J. Boyd (Ed.), *The papers of Thomas Jefferson, November 1789 to July 1790* (Vol. 16, pp. 602–617). Princeton, NJ: Princeton University Press.

Cajori, F. (1907). *A history of elementary mathematics with hints on methods of teaching.* New York, NY: Macmillan.

Cobb, L. (1835). *Cobb's ciphering book, No. 1, containing all the sums and questions for theoretical and practical exercises in Cobb's explanatory arithmetick No. 1.* Elmira, NY; Philadelphia, PA: Birdsall & Co; James Kay Jn & Brother.

Cocker, E. (1685). *Cocker's decimal arithmetick, ...* London, UK: J. Richardson.

Cocker, E. (1719). *Cocker's arithmetic: Being a plain and familiar method suitable to the meanest capacity ...* London, UK: H. Tracey.

Cocker, E. (1738). *Cocker's arithmetick.* London, UK: A. Bettesworth & C. Hitch.

Daboll, N. (1800). *Daboll's schoolmaster's assistant: Being a plain, practical system of arithmetic; adapted to the United States.* New London, CT: Samuel Green.

Dilworth, T. (1762). *The schoolmaster's assistant. Being a compendium of arithmetic, both practical and theoretical* (11th ed.). London, UK: Henry Kent.

Dilworth, T. (1773). *The schoolmasters assistant: Being a compendium of arithmetic, both practical and theoretical* (17th ed.). Philadelphia, PA: Joseph Cruikshank.

Dilworth, T. (1784). *The young book-keeper's assistant: Shewing him, in the most plain and easy manner, the Italian way of stating debtor and creditor* (9th ed.). London, UK: Richard and Henry Causton.

Dilworth, T. (1790). *The schoolmaster's assistant: Being a compendium of arithmetic both practical and theoretical.* Philadelphia, PA: Joseph Crukshank.

Dilworth, T. (1797). *The schoolmaster's assistant: Being a complete system of practical arithmetic.* Philadelphia, PA: Joseph Crukshank.

Dilworth, T. (1802). *The schoolmaster's assistant: Being a compendium of arithmetic both practical and theoretical.* Philadelphia, PA: Joseph Crukshank.

Ellerton, N. F., & Clements, M. A. (2012). *Rewriting the history of school mathematics in North America 1607–1861.* New York, NY: Springer.

Ellerton, N. F., & Clements, M. A. (2013, September). *The mathematics of decimal fractions and their introduction into British and North America schools.* Paper presented to the Third International Conference on the History of Mathematics Education, held at Uppsala University Sweden.

Ellerton, N. F., & Clements, M. A. (2014). *Abraham Lincoln's cyphering book and ten other extraordinary cyphering books.* New York, NY: Springer.

Fanning, D. F., & Newman, E. P. (2011, July 24). More on the *American Accomptant* and the first printed dollar sign. The *E-Sylum, 14.* Numosmatic Biblomania Society.

George Washington to Nicolas Pike, June 20, *1788* (in Electronic Text Center, University of Virginia Library). Retrieved December 7, 2006, from http:etext.virginia.edu/etcbin/ toccer new2?id=WasFi30.xml&ima

Goodwin, J. (2003). *Greenback: The almighty dollar and the invention of America.* New York, NY: Henry Holt and Company.

Jefferson, Thomas to Edward Carrington, May 27, 1788 (in H. A. Washington (Ed.), *The writings of Thomas Jefferson,* New York, NY: H. W. Derby, 1861).

Jefferson, T. (1785). *Notes on the establishment of a money unit and of a coinage for the United States.* Paris, France: Author. (The notes are reproduced in P. F. Ford (Ed.), *The works of Thomas Jefferson* (Vol. 4, pp. 297–313). New York, NY: G. P. Putnam's Sons. This was published in 1904).

Jess, Z. (1799). *The American tutor's assistant* ... Wilmington, DE: Bonsal and Niles.

Karpinski, L. C. (1980). *Bibliography of mathematical works printed in America through 1850.* New York, NY: Arno Press.

Lee, C. (1797). *The American accomptant; being a plain, practical and systematic compendium of Federal arithmetic* ... Lansingburgh, NY: William W. Wands.

Linklater, A. (2003). *Measuring America: How the United States was shaped by the greatest land sale in history.* New York, NY: Plume.

Macintyre, S., & Clark, A. (2004). *The history wars.* Melbourne, Australia: Melbourne University Press.

Martin, G. H. (1897). *The evolution of the Massachusetts public school system: A historical sketch.* New York, NY: D. Appleton and Company.

McCusker, J. J. (1992). *Money and exchange in Europe and America 1600–1775,* Chapel Hill, NC: University of North Carolina Press.

McCusker, J. J. (2001). *How much is that in real money? A historical commodity price index for use as a deflator of money values in the economy of the United States,* New Castle, DE: Oak Knoll Press.

Mehl, R. M. (1933). *The Star rare coin encyclopedia and premium catalog.* Fort Worth, TX: Numismatic Co. of Texas.

Monroe, W. S. (1917). *Development of arithmetic as a school subject.* Washington, DC: Government Printing Office.

Pike, N. (1788). *A new and complete system of arithmetic, composed for the use of the citizens of the United States.* Newburyport, MA: John Mycall.

Pike, N. (1793). *Abridgment for the new and complete system on arithmetic, composed for the use and adapted to the commerce of the citizens of the United States.* Newburyport, MA: Isaiah Thomas.

Pike, N. (1797). *A new and complete system of arithmetic, composed for the use of the citizens of the United States* (2nd ed.). Worcester, MA: John Mycall.

Robinson, H. (1870). *The progressive higher arithmetic for schools, academies, and mercantile colleges.* New York, NY: Ivison, Blakeman, Taylor & Co.

Root, E. (1795). *An introduction to arithmetic for the use of common schools.* Norwich, CT: Thomas Hubbard.

Simons, L. G. (1924). *Introduction of algebra into American schools in the 18th century.* Washington, DC: Department of the Interior Bureau of Education.

Skemp, R. (1976). Instrumental understanding and relational understanding. *Mathematics Teaching, 77,* 20–26.

Sprague, W. B., & Lathrop, L. E. (1857). Chauncey Lee, D.D. In W. B. Sprague, *Annals of the American pulpit: Trinitarian Congregational: Or commemorative notices of distinguished American clergymen of various denominations, from the early settlement of the country to the close of the year 1855, with historical introductions* (Vol. 2, pp. 288–291). New York, NY: Robert Carter and Brothers.

Sterry, C., & Sterry, J. (1790). *The American youth: Being a new and complete course of introductory mathematics, designed for the use of private students.* Providence, RI: Authors.

Sterry, C., & Sterry, J. (1795). *A complete exercise book in arithmetic, designed for the use of schools in the United States.* Norwich, CT: John Sterry & Co.

Tharp, P. (1798). *A new and complete system of Federal arithmetic.* Newburgh, NY: D. Denniston.

Thomas, I. & Andrews, E. J. (1809). *Preface to the third octavo edition of A new and complete edition of arithmetick, composed for the use of citizens of the United States: Seventh edition for the use of schools.* Boston, MA: Author.

Todd, J., Jess, Z., Waring, W., & Paul, J. (1800). *The American tutor's assistant; or, a compendious system of practical arithmetic; …* Philadelphia, PA: Zachariah Polson, Junior.

Walsh, M. (1801). *A new system of mercantile arithmetic: Adapted to the commerce of the United States, in its domestic and foreign relations; with forms of accounts, and other writings usually occurring trade.* Newburyport, MA: Edmund M. Blunt.

Windschuttle, K. (1996). *The killing of history: How literary critics and social theorists are murdering our past.* San Francisco, CA: Encounter Books.

Workman, B. (1789). *The American accountant or schoolmaster's new assistant …* Philadelphia, PA: John M'Culloch.

Workman, B. (1793). *The American accountant or schoolmaster's new assistant … Revised and corrected by Robert Patterson.* Philadelphia, PA: W. Young.

Chapter 6

Decimal Fractions in School Arithmetic in Great Britain and North America During the Eighteenth and Nineteenth Centuries

Abstract: This chapter compares the extent to which decimal fractions were part of author-intended and teacher-implemented school mathematics curricula in Great Britain and in North America during the eighteenth and nineteenth centuries. It is assumed that what authors of arithmetic textbooks included in their textbooks was what they hoped, and *intended*, students would study; and that what teachers required their students to write in cyphering books constituted the *implemented* arithmetic curricula for those students. Entries in 472 cyphering books—370 prepared in North America and 102 in Great Britain—were analyzed and it was concluded that, before 1792, proportionally more students in British schools studied decimal fractions than in North American schools. That was an unexpected finding given that most eighteenth-century arithmetic textbooks in both nations included sections on both vulgar and decimal fractions. Analysis also revealed that in the nineteenth century the proportion of North American students who studied decimal fractions at schools increased but in Great British it did not. The different levels of emphasis on decimal fractions in the two nations are explained through the theoretical lens of lag time.

Keywords: Abraham Lincoln; Ciphering book; Common fractions; Cyphering book; Cyphering tradition; Decimal fractions; Federal money; Francis Walkingame; Ian Westbury; Implemented curriculum; Intended curriculum; Lag time; Thomas Jefferson; U.S. weights and measures; Vulgar fractions

Lag Time for Decimal Fractions in Great Britain and the United States of America

In earlier chapters of this book three subsets of the North American cyphering books in the Ellerton-Clements collection—specifically, those cyphering books prepared during 1729–1791, 1792–1800 and 1801–1810—were studied in detail, the aim being to identify and compare teacher-implemented curricula in North American schools during the three periods. Special emphasis was given to the place of decimal arithmetic in implemented curricula. The Ellerton-Clements collection is sufficiently large that a similar method of analysis could have been carried out with subsets of cyphering books covering later periods. However, a different type of analysis, also based on cyphering books in the Ellerton-Clements collection, is presented in this penultimate chapter.

In the 1790s, after the U.S.'s decision to introduce a form of decimal currency as "Federal money," Congress declined to accept legislation which would have introduced the world's first fully decimalized system of weights and measures. In this chapter, effects on school mathematics of the decision to introduce decimal currency as the official Federal currency, yet retain the largely British, non-decimalized system of weights and measures, will be explored. The method used will be to compare the emphases on decimal fractions in implemented school curricula in North America and in Great Britain before 1791, and between 1792 and 1887. These analyses will take advantage of the fact that, in addition to the

370 cyphering-book units prepared in North America between 1667 and 1861, the Ellerton-Clements collection also includes 102 cyphering-book units prepared in Great Britain between 1747 and 1887. The Ellerton-Clements collection is the only collection of cyphering books with large numbers of cyphering books from the two nations.

The decision to compare implemented school arithmetic curricula for the two nations arose from the lag-time theoretical lens being employed for the study. That theory suggests that a period of maladjustment occurs when, and as, an established culture struggles to adapt to new economic, philosophical or material conditions. In an established culture people create and give meaning to many unique physical and mental objects. A mental object—such as the concept of a decimal fraction—becomes accepted more quickly within a culture once an influential group, not necessarily a majority of the people, has given meaning to that object. Thus, in the United States of America from 1792 onward, it is reasonable to conjecture that the existence of an officially decimalized form of Federal money gave meaning to what previously had been a barely recognized set of decimal concepts.

Ogburn (1922) posited three stages of cultural development beyond what he called the "invention" stage—accumulation, diffusion, and adjustment. In the context of the present study, accumulation of ideas about money occurred because a new concept (decimal currency) was incorporated into teacher-implemented arithmetic curricula at the same time as the old concept (sterling currency), with its pounds, shillings, and pence, lived on, as physical objects and as images of reality. The people were familiar with the old currency, and it pleased traders and merchants for the old to continue. Diffusion occurred, with decimal concepts being forced on to the general populace by Congress's decision to adopt a decimal currency. Adjustment occurred as teachers, students, parents, traders, etc., gradually acquired an increasing awareness of the arithmetic associated with decimalization. The processes of accumulation, diffusion, and adjustment occurred in different ways, at different rates, and in different parts of the United States, and in Great Britain, because of the different histories, practices, and needs in the different places.

If the theory is applicable, then one would expect that, in the United States of America, where Congress instituted a decimalized currency as Federal money, teachers would have felt pressured to teach decimal fractions to their students; whereas, in Great Britain—where sterling currency, with its pounds, shillings, pence and farthings, remained firmly in place—there would have been no such pressure. In the context of mathematics education, lag time will vary within and between mathematics education communities, depending on a community's willingness, or lack of willingness, to include a concept or principle or skill in the implemented mathematics curriculum of schools.

Decimal Fractions in British and North American School Mathematics 1630–1792

Edmund Wingate, a mathematically competent author of early British school arithmetic textbooks, boldly used the "decimal point"—a recently-developed notation. Wingate (1624, 1630) distinguished between "natural or common arithmetick" and "artificial arithmetick" (Glaisher, 1873). With artificial arithmetick he introduced decimal fractions and logarithms, and asked students not only to use these when making calculations to expedite lengthy multiplications, or divisions, but also to apply them in everyday tasks—such as finding the area of a piece of board 10¾ feet long and 5½ feet wide, by multiplying 10.75 by 5.5 and then attaching the appropriate area unit.

Although Wingate clearly *intended* decimal fractions to become an integral part of school arithmetic, it was not clear whether that was also true of John Kersey (1683). It appears that Kersey persuaded Wingate that it would be wise to place sections on vulgar and decimal fractions after the chapters on whole numbers and their applications. This allowed British teachers to continue to follow the traditional, non-decimal sequence for elementary *abbaco* arithmetic. Of course, many students would not remain at school long enough to reach the later chapters of a textbook, and so adopting Kersey's approach virtually guaranteed that teacher-implemented curricula would be unlikely to change from pre-decimal days. Other authors adopted a similar approach. Thus, for example, Edward Cocker's (1677) first major arithmetic textbook, published by John Hawkins after Cocker's death, hardly mentioned decimal fractions, but in 1685 Hawkins published *Cocker's Decimal Arithmetick* which was intended to serve the needs of those who wanted to study arithmetic from the perspective of decimal fractions (Cocker, 1685).

In both North America and Great Britain there was a well-established tradition relating to what school arithmetic should look like, and teachers who had never studied, or taught, decimal arithmetic, were naturally reluctant to embrace methods which, they thought, were unlikely to be as efficient as those that they had used for years. Cocker's (1685) *Decimal Arithmetick* did not sell nearly as well as his earlier, traditional arithmetic text. In North America the most popular arithmetic textbook during the period 1770–1800 was, easily, Thomas Dilworth's. But Dilworth—like most other authors of popular British arithmetics—left any consideration of vulgar or decimal fractions until after readers had completed the standard arithmetic curriculum using traditional methods that did not utilize vulgar or decimal fractions (see, e.g., Dilworth, 1773). This resulted in many teachers not feeling any need to change their ways. Teachers were not trained to teach, and did not have opportunities to attend professional development programs. They set out to emulate what they had done, and how they had been taught, when they themselves had been students at school—and as a result, a teacher often based his or her approach to teaching arithmetic around the well-used, and much cherished, cyphering book that he or she had prepared at school (Ellerton & Clements, 2012).

Francis Walkingame's (1785) *The Tutor's Assistant; being a Compendium of Arithmetic and a Complete Question* Book was easily the most popular school arithmetic textbook in Great Britain between 1750 and 1850 (Denniss, 2012; Michael, 1993; Stedall, 2012). In Great Britain, unlike the situation in North America, most of the students who prepared cyphering books copied their notes and examples from textbooks (Denniss, 2012; Stedall, 2012). With one or two minor exceptions (e.g., Greenwood, 1729; Venema, 1730), between 1607 and 1776 the only mathematics textbooks available for use in the British colonies in North America were those of British origin. The best-known authors of school arithmetics were Edmund Wingate, Edward Cocker, George Fisher, Thomas Dilworth, John Hill, James Hodder, and Francis Walkingame—although Walkingame's texts were more used in Canada than in the New England colonies. Most North American schools did not require students to own an arithmetic textbook and, generally speaking, students studying arithmetic in North American schools did not have access to an arithmetic textbook (Ellerton & Clements, 2012). Nevertheless, many of the teachers who taught in North America had emigrated from England, Scotland, Wales or Ireland, and therefore one might expect that the intended and implemented curricula for school arithmetic in the two nations would be similar.

The circumstances surrounding arithmetic education in schools in Great Britain were similar in some ways, but different in others, from circumstances in colonial North America. The cyphering tradition controlled school mathematics in both nations. However, whereas the population of Great Britain and Ireland in 1800 was about 15 million (Buer, 1926), that of the United States of America was only about 5.3 million, of whom almost 900000 were slaves (Dollarhide, 2001). In Great Britain in 1800 the population density was about 118 people per square mile, but in the United States it was a mere 6 people per square mile. The largest city in the United States at the time was New York, with about 60 000 people, but London had a population of about 1 million. Great Britain was in the throes of an industrial revolution, but the United States' economic position was largely based on export money derived from agrarian products such as tobacco. And, of course, although facilities for education at all levels were much more firmly established in some parts of Great Britain than in other parts of the same nation, the poverty to be found in some of the large British cities meant that many young children were sent to work in factories without having the opportunity to attend school. Because in- and between-nation differences were so great it would be dangerous to generalize about, or compare, the state of mathematics education in the two nations. That is why implemented arithmetic curricula data will be carefully examined, for if significant differences did exist, not only with respect to what was taught in the name of arithmetic in the schools but also to how it was taught, then one would expect such differences to influence lag time.

Decimal Fractions in Implemented Arithmetic Curricula in Schools in Great Britain and North America During the Eighteenth and Nineteenth Centuries

In order to investigate questions pertaining to implemented arithmetic curricula in schools in Great Britain and North America before 1792 and from 1792 to 1887, 102 British cyphering-book units, prepared between 1749 and 1887, and 370 North American cyphering-book units, prepared between 1667 and 1861, were examined. These cyphering-book units are part of the Ellerton-Clements collection.

The term "cyphering-book unit" has been used to indicate that, in those cases for which the same person had prepared more than one cyphering book in the collection, or two or more siblings had prepared cyphering books in the collection, the set of books by the same person, or by siblings, was regarded as one cyphering-book unit. Since all of the manuscripts were purchased separately, and were prepared in many parts of Great Britain and North America, there is a sense in which the collection might legitimately be thought of as comprising units that are more or less independent of each other. The same cannot be said of the manuscripts in the two largest collections of British cyphering books—one held by the Mathematical Association and the other by John Denniss (see Denniss, 2012 for summaries of those collections)—or of cyphering books held in archives in different parts of the United States of America. Note that analysis of cyphering books in the Ellerton-Clements collection has suggested that those who prepared them were mainly from middle-class families. About 80 percent of the students were male, and the cyphering books originated in many different regions in Great Britain.

Although it might seem reasonable to assume, then, that an analysis of the extent to which entries on decimal fractions can be found in the 102 British and 370 North American cyphering-book units in the Ellerton-Clements collection would permit robust generalizations to be made with respect to the implemented arithmetic curricula in British and North American schools during the eighteenth and nineteenth centuries, there are two important caveats:

1. The dates for which the cyphering books were prepared are not distributed evenly across the eighteenth and nineteenth centuries, there being much greater numbers of manuscripts for the period 1792–1887 than for the period 1667–1791.
2. The oldest of the British manuscripts in the Ellerton-Clements collection carries the date 1749, and so generalizations to the first half of the eighteenth century could not be legitimately made.

Table 6.1 was constructed with the following four questions in mind:

1. Before 1792, to what extent did (a) British, and (b) North American school students who prepared cyphering books study common (or vulgar) fractions?
2. Between 1792 and 1887, to what extent did (a) British, and (b) North American school students who prepared cyphering books study common (or vulgar) fractions?
3. Before 1792, to what extent did (a) British, and (b) North American school students who prepared cyphering books study decimal fractions?
4. Between 1792 and 1887, to what extent did (a) British, and (b) North American school students who prepared cyphering books study decimal fractions?

Table 6.1 takes account of 10 British cyphering-book units prepared between 1667 and 1791. In fact, as already stated above, the earliest British cyphering book in the collection was prepared in 1749. Each of the other 92 British cyphering-book units was prepared between 1792 and 1887. So far as the 370 North American cyphering-book units were concerned, 35 were prepared before 1792 and 335 from 1792 onward. Note that in Table 6.1 the term "cyphering-book unit" is abbreviated to "CBU," and that abbreviation will continue to be used in the remainder of this chapter.

Table 6.1

Vulgar and Decimal Fractions in British and North American Cyphering-Book Units (CBUs) 1700–1791 and 1792–1887 (% of CBUs in that Category)

Category	Presence in CBUs of sections on vulgar fractions or decimal fractions	Great Britain 1667–1791 (Percentages based on 10 CBUs)	Great Britain 1792–1887 (Percentages based on 92 CBUs)	North America 1667–1791 (Percentages based on 35 CBUs)	North America 1792–1887 (Percentages based on 335 CBUs)
A	None (algebra cyphering books)	10% (1 book)	9% (8 books)	3% (1 book)	4% (15 books)
B	Totally integrated presence (mensuration, trigonometry, surveying CBUs)	20% (2 books)	16% (15 books)	14% (5 books)	7% (25 books)
C	*Neither* vulgar fractions *nor* decimal fractions	30% (3 books)	20% (18 books)	43% (15 books)	33% (109 books)
D	Vulgar fractions, *but not* decimal fractions	20% (2 books)	39% (36 books)	29% (10 books)	13% (42 books)
E	Decimal fractions *but not* vulgar fractions	0% (No book)	0% (No book)	0% (No book)	3% (12 books)
F	*Both* decimal fractions *and* vulgar fractions	20% (2 books)	16% (15 books)	11% (4 books)	39% (132 books)

For the purpose of analysis, the CBUs were placed in six categories (see the first column of Table 6.1):

Category *A*: This category comprised CBUs which focused on algebra—and with these "algebra CBUs" there was hardly any material directly related to common or decimal fractions;

Category *B*: This category comprised CBUs dedicated to the study of at least one of mensuration, trigonometry, surveying, navigation, or geometry, and in these CBUs decimal fractions usually played an important role;

Category *C*: The third category comprised arithmetic CBUs in which there was no mention of either vulgar or decimal fractions;

Category *D*: The fourth category comprised CBUs which mentioned vulgar fractions but *not* decimal fractions;

Category *E*: CBUs in this fifth category dealt with decimal fractions, but *not* vulgar fractions;

Category *F*: CBUs in this sixth category included *both* vulgar fractions and decimal fractions.

Learning to Use Federal Money Without Learning About Decimal Fractions

The percentages shown in Table 6.1 have been rounded to the nearest whole numbers. Note that in Table 6.1 the fact that a student included pages in his or her cyphering book on decimal currency was not sufficient to justify the decision that decimal fractions were part of that student's implemented arithmetic curriculum. For decimal fractions to be regarded as part of the implemented curriculum it was necessary for the heading "Decimal Fractions" to appear, followed by a general discussion of the idea of a decimal fraction, in the cyphering book. Figure 6.1 shows the first page dealing with Federal money in Salome Russell's (1829–1833) cyphering book. Salome, who prepared her 42-page cyphering book in New London, Connecticut, did not have any preliminary pages on numeration or addition—she went straight to subtraction, and on the first page she wrote down the "equal additions" algorithm (Ellerton & Clements, 2012). Then after practicing that algorithm for almost two pages, she made the calligraphic heading "Subtraction of Federal Money," which was followed by the rule—"Place the numbers according to their value; that is, dollars under dollars, dimes under dimes, cents under cents, &c, and subtract as in whole numbers."

On the next six pages of her cyphering book Salome gave rules for multiplication of Federal money, responded to exercises, and solved elementary problems such as "What is the amount of 500 lbs of hog's lard at 15 cents per lb?" For this "hog's lard" problem, Salome wrote, as her solution:

$$\begin{array}{r} 15 \\ \underline{500} \\ \underline{75} \quad \text{Ans. 75} \end{array}$$

Then followed pages on "Division of Whole Numbers," but these did not include questions on division of Federal money. Toward the end of her book, Salome included a page headed "Examples in Reduction of Federal Money." The following rule was given:

Here I multiply by 100, the cents in a dollar; but dollars are readily brought into cents by annexing two cyphers, and into mills by annexing three cyphers. Also, any sum in Federal money may be written down as a whole number and expressed in its lowest denomination, for when dollars and cents are joined together as a whole number, without a separatrix, they will show how many cents the given number contains, and when dollars, cents and mills are so joined together they will show the whole number of mills in the given sum. Hence, properly speaking, there is no reduction of this money; for cents are readily turned into dollars by cutting off the two right-hand figures, and mills by pointing off these figures without a dot are dollars; and the figures cut off as cents, or cents and mills.

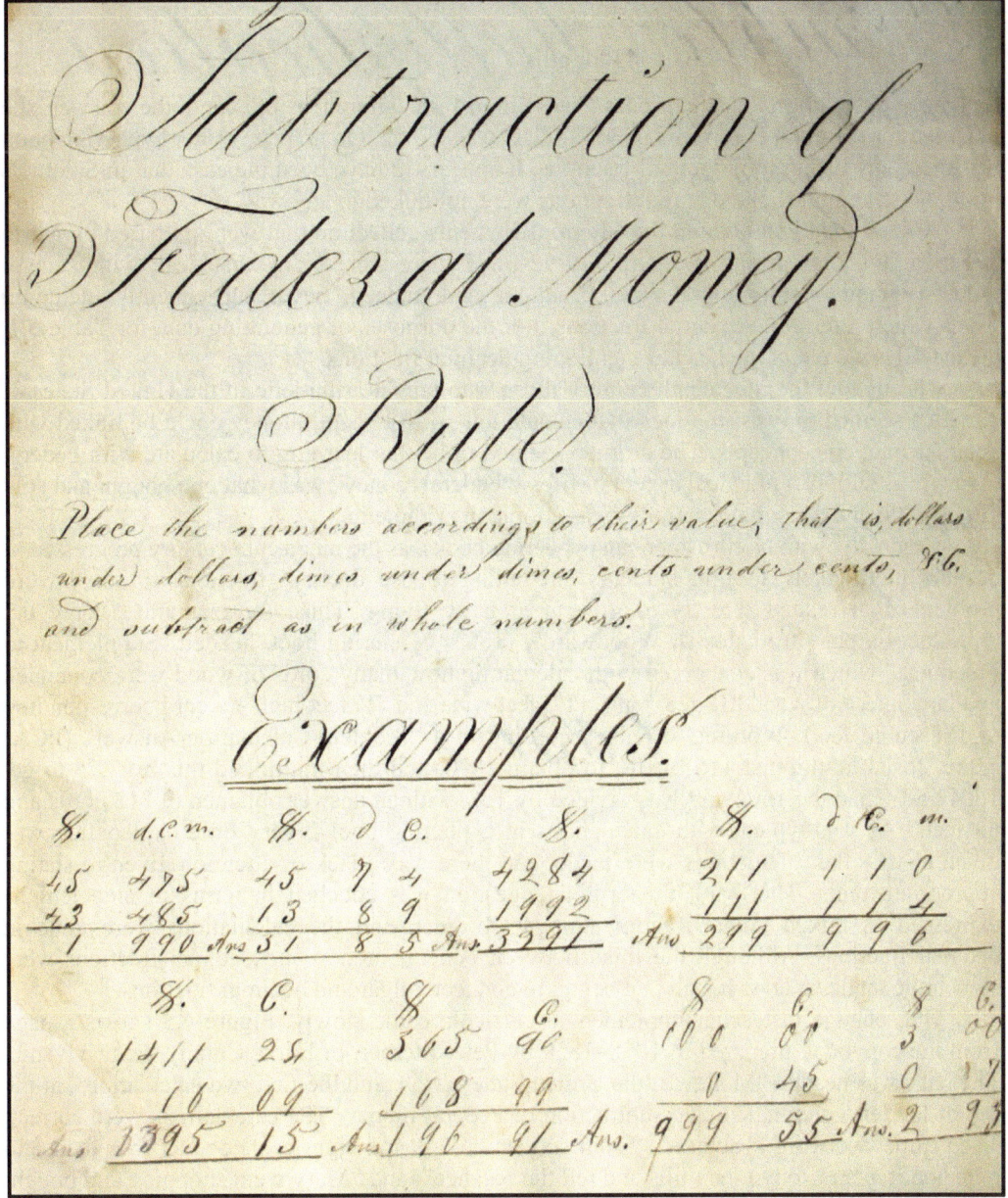

Figure 6.1. Salome Russell's (1829–1833) first cyphering-book entry on Federal money.

Then followed an illustrative example: "In 345 dollars, how may cents and mills?" For this, Salome wrote:

$$345$$

$$\underline{100}$$

$$34500 \text{ cts}$$

$$\underline{100}$$

$$\underline{34500 \text{ mills}}$$

Salome's description of the process was almost incomprehensible, and the answer she obtained for the number of mills in 345 dollars was wrong. On no page in her cyphering book was there any mention of decimal fractions. It appears to have been the case that in Salome's mind, Federal money and decimal fractions were not linked in any way.

Among the 335 CBUs in the Ellerton-Clements collection that were prepared by North American students during the period 1792–1887 there were many that dealt with Federal money without mentioning decimal fractions and, indeed, never subsequently including pages which referred to decimal fractions. For the purposes of generating data for Table 6.1, those CBUs were regarded as not mentioning decimal fractions.

Throughout the nineteenth century those who taught arithmetic in the United States of America seemed to resist the idea that calculations with Federal money could be linked with decimal fractions concepts. The attitude seemed to be that learning to calculate with Federal money was simply a practical aspect of life—"Federal money" was what one bought and sold things with, and "decimal fractions" was a topic in arithmetic.

Gradually, with North American cyphering books, as the nineteenth century progressed, it became more likely to find students attempting to use decimal fractions to solve word problems that related directly to aspects of daily living. Thus, for example, Figure 6.2 reproduces a page from Josiah Woodward's (1823) cyphering book headed "Supplement to Fractions," which was concerned with calculating how many cords of wood were contained in a large piece of wood "18 feet long, 11½ feet wide and 7¾ feet high" (a cord corresponding to 128 cubic feet). Working strictly according to a "Problem-Calculation-Answer" (PCA) genre, Josiah multiplied 115 by 18, to obtain 2070, and then multiplied this by 7.75 to get 16049,50. Then the 16049,50 was divided by 128, with an answer obtained of "12 cords and 68 feet." As was typical with calculations in cyphering books, very little explanation was given to why the calculations were made, and there was a lack of attention given to stating appropriate units. That said, the original question was couched in terms of side lengths expressed as mixed vulgar-fraction amounts of feet, and the calculations were done in decimal fractions. Although the links between Federal money and decimal fractions were slow to be made, at least Josiah had begun to connect vulgar and decimal fractions.

But, change in teacher-implemented curricula came slowly. Figure 6.3 shows a page from the copybook prepared in 1858–1859 by Peter Sterner, of Berks County, Pennsylvania. The page was headed "Money of the United State" (*sic*.), and the first two calculations, at the top of the left column, showed subtraction of Federal money. The calculations were correct. Then came two word problems: "From sixteen dollars and eight cents deduct four thousand nine hundred and sixty nine mills and tell the balance"; and "Mary went shopping and bought articles to the amount of twelve dollars and eight cents. She gave the storekeeper a twenty dollar note what change did he give her supposing him to be honest?"

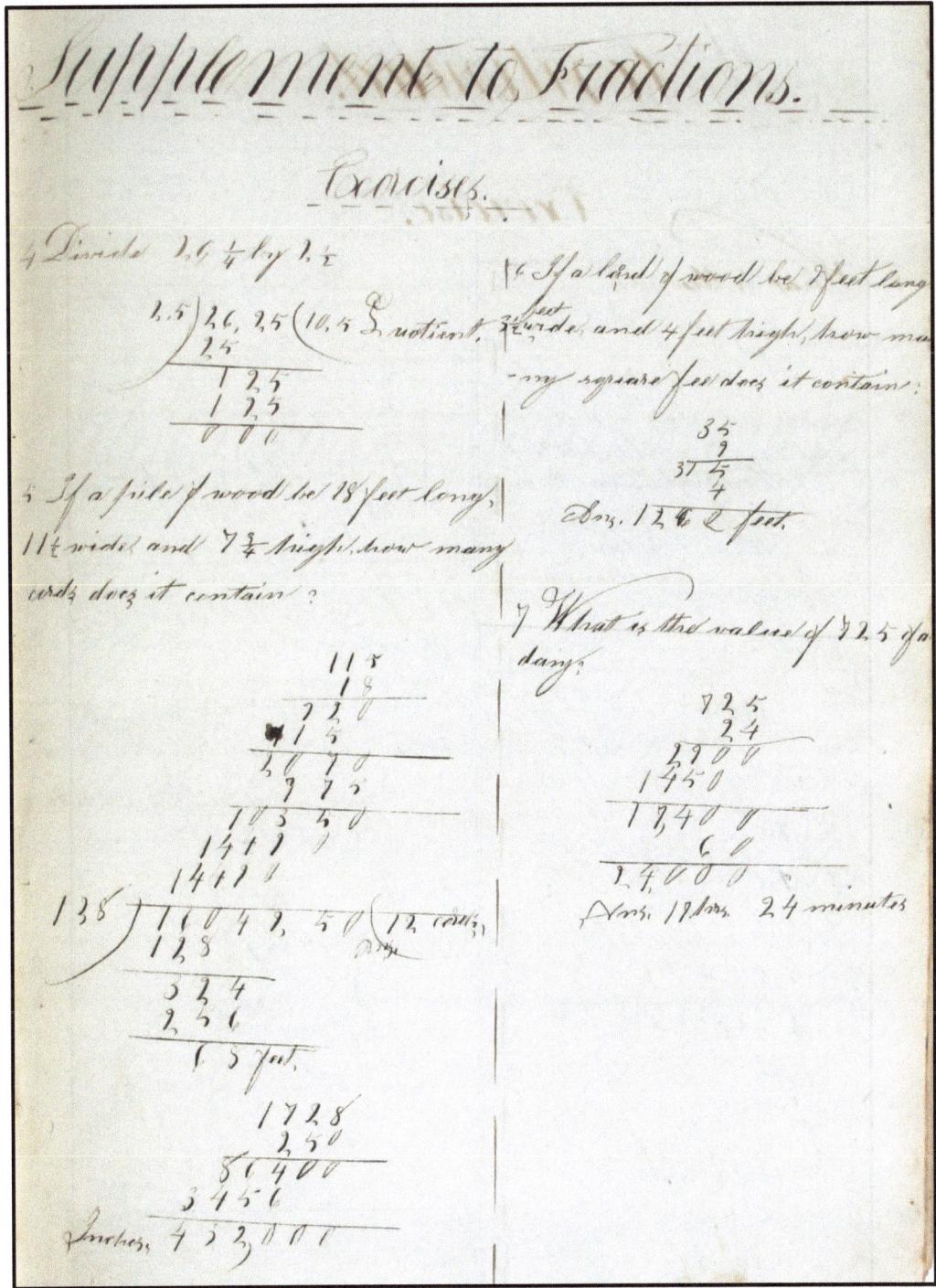

Figure 6.2. Josiah Woodward (1823) shows how vulgar and decimal fractions might be linked.

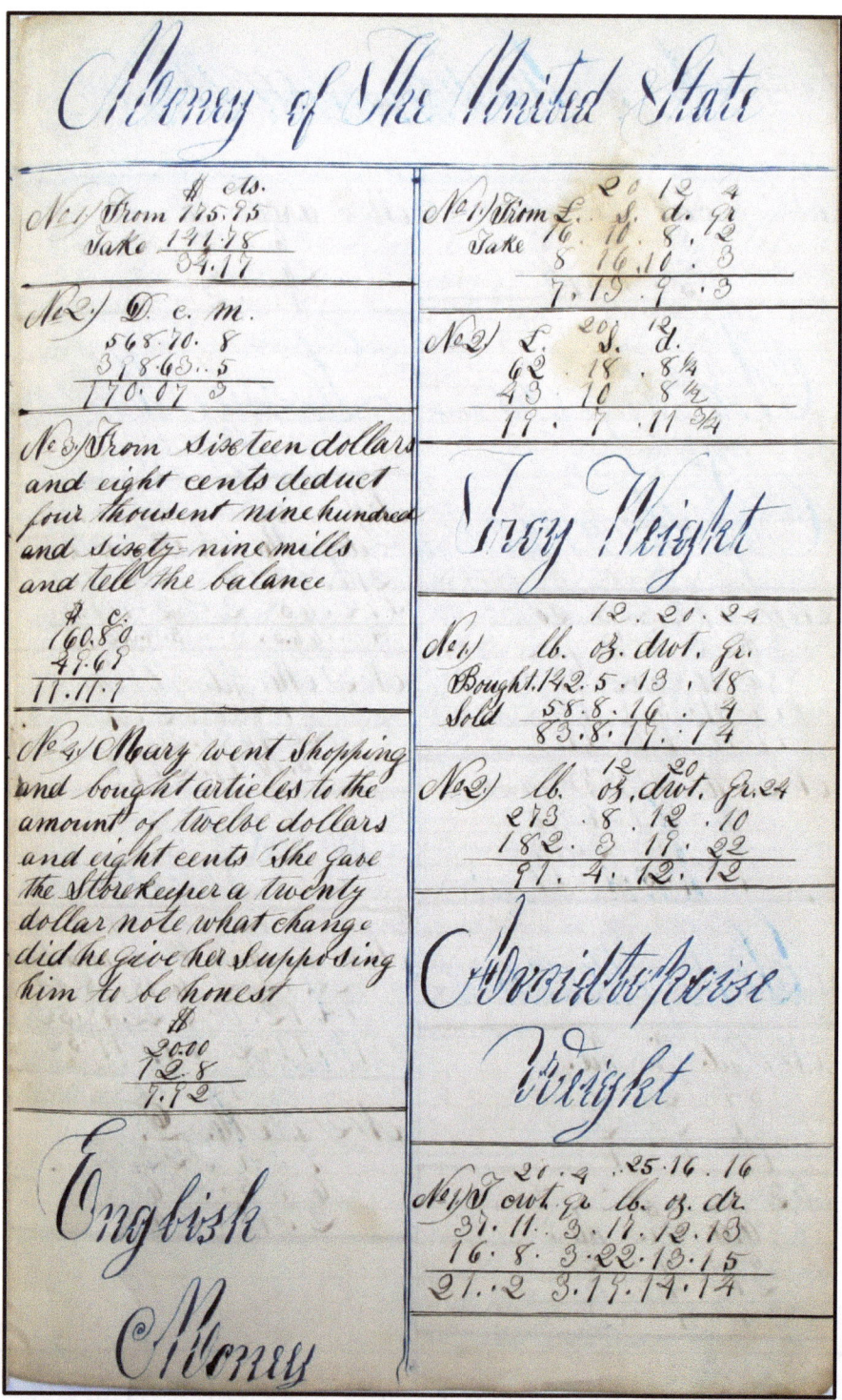

Figure 6.3. A page from Peter Sterner's (1858–1859) cyphering book.

The working for both of these word problems gave evidence of misconceptions—with the first problem, Peter apparently did not know how to write "sixteen dollars eight cents" as a decimal and he did not realize that 4969 mills was 4 dollars 96 cents and 9 mills. Peter subtracted $49.69 from $160.80. Interestingly the answer Peter gave to the question, 11.11.1, was correct—presuming he meant 11 dollars, 11 cents and 1 mill—but Peter's working indicated that he did not know what he was doing. That Peter did not know how to represent Federal money in decimal form was confirmed in his working for the second word problem, for which he wrote twelve dollars and eight cents as $12.8. Once again his answer, S7.92, was correct, but it did not correspond to the decimal subtraction that he had written down.

Immediately after attempting to solve the problems on Federal money, Peter wrote, in calligraphic form, the heading "English Money." Peter's entries were being made in his cyphering book 65 years after the first Federal coins had been issued, but pounds, shillings, pence and farthings were still being used right across the United States, and therefore teachers still felt the need to get their students to learn to operate with sterling currency. In fact, Peter's answers for the two subtraction tasks on English money were correct. On the same page were exercises involving subtraction with "Troy Weight" and "Avoidtopoise (*sic.*) Weight."

There were many pages in other cyphering books, prepared in various parts of the United States during the period 1840 through 1861, on which calculations with sterling currency units were shown. The lesson, from a historical, lag-time, point of view, is that it takes a long time to change implemented curricula—not only do the hearts, minds, and daily work practices of teachers have to be changed, but also those of parents and those operating businesses.

Readers living in the twenty-first century may find it hard to appreciate that encouraging students not only to learn decimal fractions, but also to apply the concepts in relevant situations in real life, was something which was difficult to achieve. The acceptance of decimal fractions into the implemented curriculum proved to be one of the most important developments in school mathematics curricula during the eighteenth and nineteenth centuries in North America.

Discussion

From a twenty-first century perspective, it is easy to recognize that yesterday's logarithm tables and today's inexpensive electronic calculators changed the face of calculation in contemporary school arithmetic. Thomas Jefferson may not have foreseen electronic calculators, but in the mid-1770s he knew that the kinds of calculations that ordinary people needed to do in order to survive with dignity would be learned, and carried out with much less fuss, by all who studied arithmetic, if Congress decided to introduce decimalized forms of Federal currency and weights and measures.

A strong theory should be predictive, and in this case, before the analysis which generated Table 6.1 was carried out, it was predicted that before 1792 lag time for the introduction of decimal fractions into school mathematics would be similar in both Great Britain and North America, but between 1792 and 1887 the implemented arithmetic curricula in North America would steadily have a greater emphasis on decimal fractions than implemented arithmetic curricula in Great Britain. The reason for such a prediction was that after 1792, in North America, the national currency was officially fully decimalized, with its dollars and cents, but in Great Britain the official currency, with its pounds, shillings and pence, remained non-decimalized.

The prediction that after 1791 implemented arithmetic curricula in North American schools would tend to pay greater attention to decimal fractions than was the case in British schools proved to be correct. If it is assumed that entries in Categories *B, E,* and *F* in Table 6.1 can be associated with decimal fractions being part of implemented curricula, then the sums of the percentages for those three categories for the period 1667–1791 suggest that decimal fractions were more studied in Great Britain (40% for Great Britain, and 25% for North America). But, for the period 1792–1887, the direction of the difference was reversed (32% for Great Britain, and 50% for North America). In other words, although decimal fractions came to be studied by greater percentages of school students in North America that was not the case in Great Britain—at least from a percentage-usage point of view.

The lag-time theoretical position predicted that there would be a move towards the inclusion of decimal fractions in implemented curricula in North American schools. Long before 1792 the work of research mathematicians with respect to the development of decimal concepts which would be relevant to schools had been completed. Over the next century, the potential for school education was relatively quickly translated into sections in school arithmetic textbooks (representing author-intended curricula). However, the actual teaching and learning of decimal fractions was a long time coming in schools–that is to say, decimal fractions were slow to become part of the teacher-implemented curricula in most schools. There seemed to be a need for some kind of impetus that would persuade more teachers to include decimal fractions in their implemented curricula. In North America in the 1770s and 1780s that impetus came in the form of Thomas Jefferson's (1784) surprising, brilliant, and successful push for a form of Federal currency that would be more likely to be accessible to anyone wishing to carry out everyday monetary calculations. Once Federal money had been constitutionally established, and once—after 1792—the U.S. Mint was established and was minting new coins, there were political, economic and social factors which made it difficult for schools to avoid teaching decimal fractions. Almost everyone would have thought it desirable that children learn to make Federal money calculations confidently and accurately.

In Great Britain, throughout the nineteenth century, societal pressures associated with the teaching and learning of decimal fractions remained much less compelling than was the case in the United States of America. Although, around 1790, some influential individuals in Great Britain wished to join with France and with the United States of America in introducing an international decimalized system of weights and measures (Alder, 2002), the British parliament and the U.S. Congress rejected the move and only France proceeded to decimalize weights and measures. It was not surprising that the British parliament was reluctant to consider seriously a measure that would have linked Great Britain with the French—the "old enemy"—or with the rebellious North Americans, in adopting a decimalized scheme of weights and measures. The result was that both Great Britain and the United States retained their traditional, but somewhat jumbled, systems of weights and measures. In the United States of America the decision to adopt a decimal currency as Federal money provided a fillip for the teaching and learning of decimal fractions in schools, but those teaching arithmetic in Britain schools lacked any such stimulus.

By examining entries in Categories *B, D*, and *F* in Table 6.1 it can be seen that during the period 1792–1887 more attention was paid to vulgar fractions in both North American and British schools than had been the case between 1667 and 1791. About 60% of North American students, and about 70% of British students who prepared cyphering books between 1792 and 1887 included entries on vulgar fractions (see entries in Table 6.1 for Categories *D*, and F). However, a greater proportion of students in North America than in Great Britain began to study *both* vulgar and decimal fractions.

Yet, some of the most popular U.S. school arithmetic in the first half of the nineteenth century (e.g., those authored by Daniel Adams, Nathan Daboll, Zachariah Jess, Michael Walsh, Charles Davies, Frederick Emerson, and Stephen Pike—see Ellerton and Clements, 2012) delayed the introduction of vulgar or decimal fractions, or the concept of percentage, until after they had applied "whole-number" rules in order to solve word problems associated with reduction, the rules of three (direct, inverse, and double), simple and compound interest, tare and tret, alligation, fellowship, and false position. Typically, the textbook authors included addition, subtraction, multiplication, division, and reduction of U.S. decimal currency quite early in the books, but cyphering-book evidence indicates that students approached these topics mechanically, employing rules that did not link what was being done with the mathematical, place-value concepts needed for a full appreciation of decimal fractions. When, after the four operations on whole numbers, and compound operations on quantities, had been covered, an author came to consider some of the applied topics, that author was likely to resort to presenting convoluted whole-number methods that avoided vulgar and decimal fractions. Finding solutions to these "applied" topics could have been much more straightforward if vulgar or decimal fractions concepts had been employed from the outset. One consequence of this common failure to link decimal currency with the mathematics of decimal fractions was that many teachers did not learn much about vulgar and decimal fractions themselves. The teachers encouraged their students to keep using the convoluted whole-number methods, suggested in the textbooks, to solve applied problems.

Imagine, for a moment, a student being asked to find the simple interest on 855 pounds 17 shillings and 6 pence for one year at $5\frac{3}{4}$ percent per annum, and that this student lived in a nation where dollars and cents, as well as pounds and shillings, were used in monetary transactions. Imagine, further, that this student had never formally studied percentage, or vulgar or decimal fractions. In fact, in 1825, in Pigeon Creek, a remote village in the State of Indiana in the United States of America, a tall gangling 16-year-old found himself in that situation. The youth's name was Abraham Lincoln, and at the time Abraham had never studied percentage or vulgar or decimal—see Ellerton, Aguirre Holguín and Clements (2014, pp. 166–167) for details.

Ellerton et al. (2014) reported that Abraham solved the problem using the following rule-based approach:

> Abraham first multiplied the principal (£855-17-6) by 5 (to get £4279-7-6), then he halved the principal (he erroneously obtained £427-13-6, instead of £427-18-9) and then halved what he obtained (he erroneously got £213-16-3, instead of £213-16-9). He then added his subtotals and got 49|20-17-3, inserting a vertical stroke after the "49" to indicate that 49 pounds were in the answer. From a modern perspective, the vertical stroke was used to indicate what would occur after division by 100 (for, the problem was about percent). But, what was Abraham to do with the "20" after the 49? He multiplied by 20 (to convert from pounds to shillings), and added in the 17 from the previous line, to get 417. He then used another vertical stroke after the 4 (4|17) to indicate that the 4 represented the number of shillings in the answer. Then followed 17 multiplied by 12 (to convert from shillings to pence), and with the 3 from two lines above now added in, he got 207. Another vertical stroke was used (this time after the 2, so 2|07), and after multiplying the 07 by 4 and getting 28, the 2 was taken to be the pence in the answer. Then Abraham stated his answer as £49-4-2. Although this was almost correct (the correct answer was £49-4-3), Abraham had made two calculation slips along the way. (p. 167)

Although the method Lincoln used was cumbersome, it was very similar to that shown by Stephen Pike (1822) in a model example on page 102 of the *Arithmetic* by that popular author. In the *abbaco* curriculum which Lincoln and his teachers followed, the section on simple interest came well before sections on vulgar and decimal fractions, or percentage. What is of particular relevance to this book is that the method used by the future president was the one recommended for simple interest problems by most eighteenth and early nineteenth-century British and North American authors of school arithmetics.

Denniss (2012) has argued that most teachers of arithmetic in British schools in the eighteenth and nineteenth centuries asked their students to copy pages directly from textbooks. But that does not imply that students copied all of the pages, or even most of the pages, in a textbook. From the pattern of entries in Table 6.1 it could be argued that during the period 1749–1887 a majority of students who prepared cyphering books in Great Britain did not study decimal fractions—yet most of the popular arithmetics (e.g., those by Walkingame and Dilworth) included at least one section, and often a number of sections, on decimal fractions. As the analysis in this chapter has suggested, the situation became less puzzling in the United States as the nineteenth century progressed—nevertheless, between 1793 and 1887 about one-third of U.S. school students who prepared cyphering books did not get to study vulgar or decimal fractions at school. Capable students like Abraham Lincoln coped when they were asked to solve problems which would have been much easier to solve if they had been introduced to methods involving vulgar and decimal fractions.

If one wishes to compare an "intended" curriculum with an "implemented" curriculum, one ought to consider *whose* intended curriculum one is talking about—the textbook author's, the school's, or the teacher's? A textbook author might have expected students to study both vulgar and decimal fractions. However, by placing the sections dealing with those topics *after* all major topics had been dealt with using whole-number methods—which was standard practice with authors of arithmetics, both in Great Britain and in North America—the authors made it easy for teachers to decide that there was no need, or time, for their students to study vulgar or decimal fractions.

In any case, some of the textbook authors who included sections in their textbooks on vulgar and decimal fractions may not have intended that all students who used their books would engage with those sections—rather, the authors' reason for including the sections may have been that any students or teachers who wished to deal with vulgar or decimal fractions would have been able to do so without having to find another book.

Given the different social, economic and education circumstances inevitably associated with different schools, one could excuse teachers who decided that their students should not be compelled to study either vulgar or decimal fractions; or, the decision might have been that one should be studied but the other omitted. Such a decision could easily have been made by teachers, and the consequences would almost certainly have remained hidden from relevant local authorities.

Thus, curricula implemented by teachers were only rarely the same as curricula intended by textbook authors. But, from 1792 onward in the United States of America the existence of a national decimalized currency placed pressure on teachers to teach their students about Federal money and therefore, indirectly, about decimal fractions. Understandably, most parents would have wanted their children to be able to make efficient calculations in Federal money, and the students themselves would have recognized that being able to perform such calculations could be useful. Teachers, too, would have recognized that if it became known that they were able to teach decimal fractions in an authoritative way then they were likely to be preferred for employment in local schools over others who did not know how to teach decimals.

From a historical perspective, the above analysis seems to call for the following four closing comments:

1. Those who research the history of mathematics curricula in Great Britain and in North America in the eighteenth and nineteenth centuries need to take account of cyphering-book data as well as the content of chapters in textbooks.
2. From a lag-time perspective, despite the pioneering efforts of scholars such as Edmund Wingate, Edward Cocker, Francis Walkingame, and Thomas Dilworth, most arithmetic teachers in British schools did not take the opportunity, which the textbook authors offered them, to teach students about decimal fractions. A majority of teachers simply did not require their students to study decimal fractions.
3. With respect to Figure 1.3, it is intriguing that the genius and hard work of research mathematicians such as François Viète, Simon Stevin, John Napier, Henry Briggs and Edmund Wingate, in introducing decimal and logarithmic concepts and showing how these could be useful to society, and even—in the case of Wingate—making efforts to import the concepts into school mathematics curricula, were not enough to encourage teachers of that period to include decimal fractions in their implemented mathematics curricula.
4. Without the fillip provided by the introduction of a decimalized Federal money in the United States, it is unlikely that the proportion of students dealing with fractions in U.S. schools would have increased as rapidly as it did in the nineteenth century.

With respect to the first of these points, this book represents, as far as we know, the first systematic attempt to take account of both intended and implemented curricula in studying the development of school mathematics in the eighteenth and nineteenth centuries. One reason for that has been that cyphering books have not been conveniently available to historians, and without cyphering-book evidence it has been impossible to provide an adequate description or analysis of implemented curricula. In recent years sizeable collections of cyphering books have become available to researchers in both Great Britain and the United States of America (see, e.g. Denniss, 2012; Ellerton & Clements, 2012). Textbooks have always been much more readily available than cyphering books, and have been much more widely consulted by historians. However, although the textbooks are likely to bear some resemblance to the authors' intended curricula, by themselves they do not enable implemented curricula to be well examined. A second reason why historians have not contrasted intended and implemented curricula is that it was not until about 1980 that distinctions between intended, implemented, and attained curricula were introduced—by Ian Westbury (1980), the editor of the *Journal of Curriculum Studies*.

With respect to the second of these points, one might ask the pointed question: With respect to decimal fractions, why should teachers in Great Britain have departed from their comfort zones? Those who taught arithmetic in British schools were familiar with the traditional arithmetic curriculum, and from their perspectives, there seemed to be no pressing reasons why they should have asked their students to study decimal fractions. Currency calculations, and calculations related to weights and measures, could be performed without using decimals—indeed, such calculations had been done without decimals for centuries. So why change?

With respect to the third point, those students in the eighteenth and nineteenth centuries who studied navigation, astronomy, mensuration, and surveying quickly began to make use of decimal and logarithmic concepts. The reason was simple—teachers and students of such subjects quickly recognized that decimals, logarithms and trigonometric concepts greatly facilitated the study of navigation, astronomy, mensuration, and surveying. The effort that teachers and students would need to make in order to become familiar with the new material was deemed to be worth it. But, despite Wingate's best efforts, a similar attitude was not to be found among teachers and students of traditional school arithmetic. That said, it should be recognized that students of navigation, astronomy, surveying, mensuration, etc., often struggled to gain relational understandings of the key decimal concepts and principles involved when they attempted to apply logarithmic theory (Ellerton & Clements, 2014).

It is likely that the situation with respect to the teaching and learning of decimals in Great Britain gradually changed after the 1860s, when an externally-set arithmetic curriculum was enforced in government primary schools, and externally-set examinations became increasingly important. The compulsory arithmetic examinations would include questions on decimal arithmetic, and a "payment-by-results" system would mean that teachers' salaries would depend on how well their students would answer such questions. That created an altogether different education scenario (for details, see Roach, 1971). Further research on the effects of the introduction of the payment-by-results system on intended, implemented, and attained mathematics curricula is needed.

In many parts of North America, although perhaps not all parts, the large-scale introduction of decimal fractions into implemented school curricula had a shorter time-lag than what occurred in Great Britain. From our analysis of data reported in this book we are confident that the shorter time-lag in parts of the United States of America was because the United States introduced decimalized currency in the 1790s, and hence knowledge of how to perform decimal calculations obviously became useful at all levels of society. But that conjecture needs further investigation, for we have found that as late as the mid-1820s many students did not learn about decimal fractions at school. Abraham Lincoln was one such student—but, undoubtedly, this state of affairs was not confined to students and teachers in Indiana.

Entries in Table 6.1 indicated that many teachers chose not to introduce decimal fractions to their students. One might conjecture that because weights and measures were not decimalized in either Great Britain or in the United States of America, decimal fractions were rarely used in either of those nations in the nineteenth century for calculations that related to weights and measures. Cyphering-book evidence suggests that that was true even for problem situations—like, for example, calculating areas of rectangular regions—in which decimalization would have been practically and educationally advantageous. From that perspective, it would be interesting to complement the entries in Table 6.1 with entries derived from cyphering books prepared in France. Although we are not well placed to conduct that research, it does need to be carried out. In the post-revolutionary and Napoleonic eras in France a more centralized arithmetic curriculum in France was established, and it is likely that that would have encouraged more secondary teachers in France to teach, and their students to learn and use, vulgar and decimal fractions than had been the case in the pre-revolutionary period (Barnard, 2008).

Although it is impossible to ascertain *attained* curricula merely by examining intended or implemented curricula, when examining cyphering books one gets a sense of the extent to which the students who prepared those books understood what they were writing about. One would like to know more about whether the students *reflected* on what they entered in their cyphering books—and, if they did, on the quality of their reflections.

In this chapter curriculum questions relating to the introduction and purpose of vulgar and decimal fractions in school mathematics curricula in North America and Great Britain in the seventeenth, eighteenth, and nineteenth centuries have been asked, and related issues examined. Emphasis has been placed not only on distinctions between the concepts of intended and implemented curricula, but also on the question of *whose* intended curricula were being considered. Using a lag-time theoretical perspective it became clear that curriculum development in school mathematics is something far more complex than merely setting out neatly those ideas which mathematicians or politicians or curriculum developers, or relevant significant others, have deemed to be important for students to learn and apply. One of the consequences of the introduction of a decimalized currency in 1792 by the fledgling United States was that more students in North America than in Great Britain would learn about decimal fractions, and about how they could be applied to solving real-life problems.

In Great Britain in the late 1850s, a report of a government Inquiry into the State of Popular Education revealed that less than 70% of children attending English public schools, and only about one-third of children attending private schools, were taught arithmetic. And, of course, many children did not attend school at all. It was not until 1871 that "proportion and vulgar and decimal fractions" were introduced into the mandated Standard VI curriculum on which payment-by-results examinations were based (Howson & Rogers, 2014, p. 258). Thus, more than 250 years after the pioneering work of Stevin, Napier, and others, decimal fractions were still not being studied by many children in England—despite the fact that authors of popular British school arithmetics had, for many years, included sections on decimal fractions. Lag time was much greater than one would have expected, and that fact has escaped the attention of historians whose focus has been on the content of mathematics textbooks. We cannot over-emphasize the importance, from a historical perspective, of our conclusion that analysis of the content of textbooks alone will not provide a reliable guide to curricula implemented in classrooms.

This chapter has offered quantitative and qualitative analyses of cyphering-book data related to the teaching and learning of vulgar and decimal fractions in North America and in Great Britain between 1667 and 1887. Our analyses were guided by the lag-time theoretical perspective. In the future we intend to undertake similar curricular analyses, in areas other than vulgar and decimal fractions.

References

Alder, K. (2002). *The measure of all things: The seven-year odyssey and hidden error that transformed the world.* New York, NY: The Free Press.

Barnard, H. C. (2008). *Education and the French Revolution.* Cambridge: Cambridge University Press.

Buer, M. C. (1926). *Wealth and population in the early days of the industrial revolution,* London, UK: George Routledge & Sons.

Cocker, E. (1677). *Cocker's Arithmetick: Being a plain and familiar method suitable to the meanest capacity for the full understanding of that incomparable art, as it is now taught by the ablest school-masters in city and country.* London, UK: John Hawkins.

Cocker, E. (1685). *Cocker's Decimal Arithmetick, Wherein is shewed the nature and use of decimal fractions, in the usual rules of arithmetick, and the mensuration of plains and solids.* ... London, UK: J. Richardson.

Denniss, J. (2012). *Figuring it out: Children's arithmetical manuscripts 1680–1880.* Oxford, UK: Huxley Scientific Press.

Dilworth, T. (1773). *The schoolmasters assistant: Being a compendium of arithmetic, both practical and theoretical* (17th ed.). Philadelphia, PA: Joseph Crukshank.

Dollarhide, W. (2001). *The census book: A genealogist's guide to Federal census facts, schedules and indexes.* North Salt Lake, UT: HeritageQuest.

Ellerton, N. F., Aguirre Holguín, V., & Clements, M. A (2014). He would be good: Abraham Lincoln's early mathematics, 1819–1826. In N. F. Ellerton & M. A. Clements (Eds.), *Abraham Lincoln's cyphering book and ten other extraordinary cyphering books* (pp. 123–186). New York, NY: Springer.

Ellerton, N. F., & Clements, M. A. (2012). *Rewriting the history of school mathematics in North America 1607–1861.* New York, NY: Springer.

Ellerton, N. F., & Clements, M. A (2014). *Abraham Lincoln's cyphering book and ten other extraordinary cyphering books.* New York, NY: Springer.

Glaisher, J. W. L. (1873). On the introduction of the decimal point into arithmetic. *Report of the Meeting of the British Association for the Advancement of Science, 43,* 13–17.

Greenwood, I. (1729). *Arithmetick, vulgar and decimal, with the application thereof to a variety of cases in trade and commerce.* Boston, MA: Kneeland & Green.

Howson, G., & Rogers, L. (2014). Mathematics education in the United Kingdom. In A. Karp & G. Schubring (Eds.), *Handbook on the history of mathematics education* (pp. 257–282). New York, NY: Springer.

Jefferson, T. (1784). *Notes on the establishment of a money unit and of a coinage for the United States* (handwritten manuscript). Washington, DC: Library of Congress.

Kersey, J. (1683). *Mr Wingate's Arithmetick, containing a plain and familiar method for attaining the knowledge and practice of common arithmetick* (8th ed.). London, United Kingdom: John Martyn.

Michael, I. (1993). The textbook as a commodity: Walkingame's *The Tutor's Assistant.* *Paradigm, 12,* 2–10.

Ogburn, W. F. (1922). *Social changes with respect to culture and original nature.* New York, NY: B. W. Huebsch.

Pike, S. (1822). *The teacher's assistant or a system of practical arithmetic; wherein the several rules of that useful science, are illustrated by a variety of examples, a large proportion of which are in Federal money. The whole is designed to abridge the labour of teachers, and to facilitate the instruction of youth.* Philadelphia, PA: Benjamin Warner.

Roach, J. (1971). *Public examinations in England 1850-1900.* Cambridge, UK: Cambridge University Press

Stedall, J. (2012). *The history of mathematics: A very short introduction.* Oxford, UK: Oxford University Press.

Venema, P. (1730). *Arithmetica of Cyffer-Konst, volgens de Munten Maten en Gewigten te Nieu-York, gebruykelyk als mede een kort Ontwerp van de Algebra.* New York, NY: Jacob Goelet.

Walkingame, F. (1785). *The tutor's assistant: Being a compendium of arithmetic and a complete question book* (21st ed.). London, UK: J. Scratcherd & I. Whitaker.

Wingate, E. (1624). *L'usage de la règle de proportion en arithmétique.* Paris, France: Author.

Wingate, E. (1630). *Of natural and artificiall arithmetique.* London, United Kingdom: Author.

Wingate, E. (1630). *Arithmétique made easie.* London, UK: Stephens and Meredith.

Chapter 7

Decimal Fractions and Curriculum Change in School Arithmetic in North America in the Eighteenth and Nineteenth Centuries

Abstract: The main purpose of this final chapter is to answer each of the five research questions identified in Chapter 1. In the eighteenth century the British government had transplanted its sterling pounds, shillings, pence and farthings firmly into the British colonies in North America and since the *abbaco* curriculum followed by those who taught school arithmetic was deeply concerned with money calculations, implemented curricula for school arithmetic in North America asked students to perform countless money calculations based on sterling currency. All of that changed after Thomas Jefferson's (1784) successful efforts to introduce a decimalized currency as Federal money, for then the new nation's schools, and teachers were called upon to re-focus their curricula so that new forms of money calculations would be included. However, the new focus was blurred by Congress's decision not to introduce a Federal decimalized system of weights and measures, and the decision to allow the states to continue to use their own versions of sterling money as "legal" currency. As a result, the shift in emphasis resulting from the courageous decision to introduce decimal currency was slow to take hold. The chapter concludes with commentary on the limitations of the study, and with the identification of researchable questions that scholars might profitably address in the future.

Keywords: Common fractions; Ciphering book; Cyphering book; Decimal currency; Decimal fractions; Federal money; History of mathematics textbooks; Implemented curriculum; Intended curriculum; Lag time; Thomas Jefferson; Vulgar fractions; U.S. weights and measures

Answering the Five Research Questions

In Chapter 1 the following five research questions were identified:

1. To what extent did policies and practices relating to money influence school arithmetic curricula in the North American colonies before 1790?
2. To what extent did policies and practices with respect to weights and measures influence school arithmetic curricula in the United States of America before 1790?
3. To what extent did the place of decimal fractions in U.S. school arithmetic curricula change as a result of the introduction of a decimalized scheme of Federal money around 1790?
4. To what extent did policies and practices with respect to weights and measures play a part in school arithmetic curricula in the United States of America between 1791 and 1860?
5. What are the implications of the findings of this study for mathematics curriculum theory?

M. A. (Ken) Clements, & N. F. Ellerton, *Thomas Jefferson and his decimals 1775–1810: Neglected years in the history of U.S. school mathematics*, DOI 10.1007/978-3-319-02505-6_7,
© Springer International Publishing Switzerland 2015

In this the final chapter these questions will be answered, one by one. Then some limitations of the study will be offered, and questions raised for further research.

Before answering the five questions it will be useful to offer brief commentary on three wider historical and educational issues which, it might be argued, have underpinned the study. The first of these issues, which seemed to demand more and more attention as the book was being written, relates to how the importance and roles of vulgar and decimal fractions in school mathematics curricula changed during the eighteenth and nineteenth centuries, especially in the British North American colonies and, subsequently, in the United States of America. The second issue is concerned with the fundamentally important role of Thomas Jefferson as a change agent for school mathematics curricula—obviously, but not only, in the United States of America. The third issue relates to the importance of the distinction between author-intended and teacher-implemented school arithmetic curricula. We now comment on those three issues, necessarily briefly.

During the Eighteenth Century most North American School Students did not Study Vulgar or Decimal Fractions

If one analyses school arithmetic textbooks used in the United States during the eighteenth and nineteenth centuries, and cyphering books prepared at the same time, one is struck by the emphasis in those books on elementary arithmetical operations, on money problems, and on problems relating to weights and measures. These emphases were part and parcel of merchant-driven curricula which influenced school mathematics in Great Britain in the seventeenth and eighteenth centuries (Ellerton & Clements, 2014), when they were transplanted into the British North American colonies (Ellerton & Clements, 2012). Part of these curricula was the rule of three, which related to problems dealing with proportional relationships—and almost always was set in contexts related to money or weights and measures, or both. Entries in Table 2.1 (which was concerned with author-intended curricula) and Table 2.2 (concerned with teacher-implemented curricula) suggest that about 60% of time spent on school mathematics in the British North American colonies in the eighteenth century was dedicated to elementary arithmetic operations or to the arithmetic of money and of weights and measures.

The importance attached to money and weights and measures needs to be understood if the study described in this book is to be fully appreciated. Any serious attempt to change the order and balance of what was done so far as numeration, the four operations, reduction, money, weights and measures, and the rules of three, is worthy of attention because if it were to be successful then such a change would alter implemented arithmetic curricula in schools in fundamental ways.

The traditional *abbaco* curriculum placed the study of vulgar fractions and decimal fractions *after* the arithmetic of whole numbers had been completed and applied though topics like the rules of practice, the rules of three, currency exchange, loss and gain, tare and tret, barter, equations of payment, alligation, fellowship, and false position. With *abbaco* curricula, once these topics had been dealt with they were then revisited, with the emphasis now being on how vulgar fractions could be used in the practical contexts. After that had been done, a second revisiting occurred, with the emphasis now being on how decimal fractions could be used in the various practical contexts. These "revisitings" found expression in the common practice, by authors of arithmetic textbooks—see, for example, those by Thomas Dilworth (1773) and Francis Walkingame (1785)—of including a section on "vulgar arithmetic" after a section on the "arithmetic of whole numbers," and then having a section on "decimal arithmetic" after the section on vulgar arithmetic (see Michael, 1993).

Persons seeking to understand arithmetic curricula in North American schools in the eighteenth and nineteenth centuries need to move their attention beyond what textbook authors wrote in their textbooks, for if they confine themselves to a consideration of textbooks they are likely to conclude that both vulgar and decimal fractions were well studied in almost all schools. The evidence presented in this book points to the opposite conclusion—that, in fact, vulgar and decimal fractions were studied by only a minority of students in a minority of schools. It is possible that most authors of arithmetic textbooks realized that in many cases—indeed, almost certainly a majority of cases—students would not proceed beyond the first, whole-number sections in their books. That whole-number arithmetic was the chief focus of attention in school arithmetic can be confirmed by studying cyphering books. While carrying out the research for this study we had daily access, over a period of years, to large collections of both arithmetic textbooks and cyphering books, and there can be little doubt that the curriculum patterns we have just described represented the reality of school arithmetic in North America in the eighteenth century and for the first half of the nineteenth century.

The Fundamentally Important Role of Thomas Jefferson as a Change Agent for School Mathematics Curricula in the United States

The title of this book begins with the words "Thomas Jefferson and his Decimals" and, in case any reader might think that those words were chosen merely to attract interest in the book, we would like to say, explicitly, that we believe that Thomas Jefferson was responsible for achieving the most important change in the implemented curricula of school mathematics in North America during the years 1790 through 1820, and perhaps well beyond that.

Jefferson was not a teacher, and it is likely that he knew little of what happened in the name of mathematics in North American schools. His own education, as the son of a wealthy planter in Virginia, had been extremely privileged, and did not provide him with knowledge of what happened in ordinary schools across North America. He developed a genuine passion for mathematics, and only dimly recognized that very few students in North America ever got to study algebra, or trigonometry, or geometry, or mensuration or calculus. In North American schools, *abbaco* arithmetic was the order of day, but it was mainly studied by boys, and almost always in a way consistent with the cyphering tradition. Cyphering books testified to implemented arithmetic curricula in schools (Ellerton & Clements, 2012). Jefferson did not know much about what happened in ordinary district schools.

What Jefferson did have, though, was an incredible mind which led him to the conclusion that, from an education perspective, it would be easier for children, and adults, to learn to read, write, speak and apply arithmetic if decimal fractions were to become a vital component of school mathematics. He also had enough political experience, and astuteness, to recognize that in 1775 the new nation had a rare opportunity to persuade citizens to think about relationships in terms of decimal quantities by introducing systems of coinage and weights and measures that were themselves fully decimalized. In the twenty-first century it is difficult to appreciate how amazing Jefferson's insights were so far as decimalization was concerned.

In his remarkable "Notes on Coinage" (Jefferson, 1784), written *before* he went to Paris Jefferson argued the case for the introduction of a fully decimalized Federal system of money. The main reason he put forward for why change was needed was an educational one—the nation needed to do this, he insisted, because it would enable citizens to perform more easily all the arithmetical operations they would have to do in order to survive with dignity in their everyday lives. What was even more remarkable was the fact that *before he went to Paris* he also argued the case for the introduction of a fully decimalized system of weights and measures that would have coordinated relationships between time, length, area, volume, capacity, angle, and weight (Boyd, 1961). The "seconds pendulum" was at the heart of his thinking about weights and measures, and it is arguable that through his conversations with Talleyrand and Condorcet, and others, in Paris, he had a large influence on France's decision to introduce its metric system in the 1790s—though the metric system did not incorporate Jefferson's beloved seconds pendulum.

Although he was supported, in his push for a decimalized system of weights and measures, by Washington, Madison, and Hamilton, Jefferson was not able to win the day politically on that issue (Boyd, 1961; Linklater, 2003). Jefferson sensed that his scheme was too radical to get past Congress, and he was right on that point. Although the refusal of Congress to accept Jefferson's recommended system for weights and measures can be interpreted as one of the most important political failures in Jefferson's career, another way of thinking would point out how that failure made clear the magnitude of his achievement in managing to win acceptance for his decimalized system of Federal money—the U.S.'s dollars and cents would become the mainstay of the most important currency system the world has ever seen. All of a sudden the potential of decimal fractions for wide-scale application in real-life situations, and not only for navigators, or surveyors or astronomers, was made plain for all to see (Dantzig, 1954). This book has been about how that system gradually welded its way into implemented school mathematics curricula in the United States of America. Paradoxically, the slowness of uptake, revealed in this book, testifies abundantly to the scope of Jefferson's achievement.

The Distinction Between Author-Intended and Teacher-Implemented School Arithmetic Curricula

Although the distinction between intended, implemented and attained curricula is of recent origin (see Westbury, 1980), it is one that has been particularly useful for the themes and issues investigated. The position taken in this book has been that, traditionally, persons investigating the history of school mathematics curricula have too often focused almost exclusively on evidence gained from textbooks or from official documents printed by colleges, schools, institutions, or systems of education. For the early history of school mathematics in North America—for the period between 1607 and 1850, say—that focus has been excusable because it was thought to be almost impossible to get a large volume of data from other sources, largely because it was thought that the "other sources" no longer existed.

But, the ferreting out and bringing together in recent years of substantial collections of students' handwritten cyphering books which were prepared by "ordinary students" in the seventeenth, eighteenth, and nineteenth centuries—for example, the Ellerton-Clements collection of over 500 cyphering books, among which 370 were prepared in North America and 102 in the United Kingdom, and the collections held by the Mathematical Association, in England, and John Denniss (2012), also in England—has stimulated some historians to seek out cyphering books among family papers in archives such as those at the Phillips Library, Salem, Massachusetts, Harvard University, Yale University, the University of North Carolina, the University of Michigan, and the College of William and Mary (Ellerton & Clements, 2012). As a result of the increasing availability of substantial numbers of cyphering books, curriculum historians have begun to reflect on the role that these manuscripts might have for curriculum history investigations in the United States of America, Great Britain, and elsewhere.

In this book it has been argued that the distinction between author-intended curricula (as can be investigated through textbooks and official documents of education institutions) and teacher-implemented curricula (as can be investigated through cyphering books) is important. As Lao Genevra Simons (1936) pointed out, evidence of implemented curricula provided by cyphering books is "unmistakable" (p. 588). It is difficult to counter an argument that what a student wrote on a daily basis in his or her cyphering book did not correspond to the implemented curriculum of that student's teacher.

When one compares teacher-intended with teacher-implemented school arithmetic curricula in North America and in Great Britain during the eighteenth and nineteenth centuries one cannot escape from the conclusion that the two versions of curricula differed in important ways. From the perspective of this book, the most important difference was that although sections on decimal fractions were almost always found in textbooks, that was not the case with the cyphering books. That observation prompted the conclusion that curricula as set out in textbooks were not more important, from a history-of-curriculum viewpoint, than what was written in cyphering books.

On reflection, we recognized that it might be misleading to label the contents of a textbook as an author-intended curriculum. The author may have placed the sections in a textbook so that the essential, "core," material came first, and the less essential, "optional," material later. This kind of thinking may have resulted in the first 100 pages or so of an arithmetic textbook dealing with the arithmetic of whole numbers and applications, and then later pages dealing with vulgar and decimal fractions. That arrangement was found in many of the textbooks studied and reported on in this book. Despite the possibility that authors did not intend readers to study carefully all of the sections in their textbooks, the term "author-intended curricula" has still been used in this book—and readers are asked to take account of the possibility that authors may not have intended that all sections would be studied by everyone who used their textbooks.

It should be noted, too, that in North America many school students who studied arithmetic in the eighteenth and early nineteenth centuries did not own an arithmetic textbook, and often their teachers did not own a textbook, either. Indeed, it was more likely that a teacher, and his or her students, would have used an existing cyphering book as the basis for what arithmetic would be studied—in other words, for a particular teacher a cyphering book, or a part of a cyphering book, could have defined the intended curriculum for one or more students in a school.

Research Question 1: To What Extent Did Policies and Practices with Respect to Money Influence School Arithmetic Curricula in the North American Colonies Before 1790?

The short answer to this question is "a great deal." British currency policies and practices were taken for granted by British authors of arithmetic textbooks, and these same policies and practices were implicitly assumed to be relevant in the schools, both in Great Britain and in North America. Evidence for the large influence can be found in every arithmetic textbook written by a British author and used in the North American colonies, and in every cyphering book in the Ellerton-Clements collection that was prepared before 1790. Coming to grips with sterling currency, and the array of relationships defining the use of weights and measures in Great Britain, took up much of the time which most students studying arithmetic in colonial North American schools were required to devote to the subject. Because the *abbaco* curriculum emerged from mercantile environments in which the solving of problems related to money and to weights and measures was centrally important, almost all of the students who "cyphered" in the schools of colonial North America grappled, on a daily basis, with tasks involving sterling currency and British weights and measures. That was what school arithmetic was all about. Entries in Tables 2.1 and 2.2 (in Chapter 2), in Table 5.1 (in Chapter 5), and in Table 6.1 (in Chapter 6) leave one in little doubt that, in fact, most of the time spent on arithmetic by school students in North America in the eighteenth century, and in fact in the nineteenth century as well, was devoted to learning content and solving word problems related to either money or to weights and measures, or to both.

Before 1775 the currencies commonly used in the various British colonies in North America were not decimalized—the currencies associated with school arithmetic were usually of the sterling variety, for which 4 farthings made 1 penny, 12 pence made a shilling and 20 shillings made 1 pound. But a shilling in one colony could have different purchasing powers from a shilling in neighboring colonies, so that if and when someone crossed a border then arithmetical confusions were likely to arise with respect to monetary transactions—even though, structurally speaking, the ratios between the values of pounds, shillings, pence and farthings within the same colony remained constant. The losers were typically "ordinary people," for businessmen and traders quickly learned how to manipulate the system. The fact of constant ratios yet, simultaneously, different boundary values, became part of what was to be learned under the topic "Exchange"—but that topic was sufficiently far along the *abbaco* sequence that many students did not reach it.

So far as weights and measures were concerned, many eighteenth-century cyphering books had page-after-page of statements showing and applying relationships between the units for measurement of troy weight, avoirdupois weight, apothecaries weight, land measure, solid measure, wine measure, beer and ale measure, cloth measure, time, etc. Relationships were copied into cyphering books, and exercise after exercise was devoted to solving simple problems involving those relationships.

But, it would be wrong to leave the impression that government policy—either colonial government policy, or British government policy—directly influenced the arithmetic curricula of schools in the colonies. Rather, the influence was indirect—but no less powerful because of that. There was no official, well-defined British national arithmetic curriculum, or colonial government-defined arithmetic curriculum. Nevertheless, sterling currency and weights and measures relationships needed to be known—even by students in families from Continental European countries (such as with many German-background families in Pennsylvania, and many Dutch-background families in and around New York).

Some historians (e.g., Cohen, 1982) have claimed that for most teachers, and students, of eighteenth-century school arithmetic in the North American colonies, the subject was all about memorization. Cyphering-book evidence suggests, however, that that generalization presents an overly negative evaluation of the cyphering tradition. School arithmetic offered plenty of opportunities for genuine problem solving—after analyzing many North American cyphering books, in the Ellerton-Clements collection and elsewhere, Ellerton and Clements (2012) referred to a "socially-oriented, structure-based, problem-solving theory of learning" (p. 143) which was behind entries in a majority of the cyphering books. Pages in most arithmetic textbooks and cyphering books showed solutions to numerous problems. Some of the problems—like those in which the measure of a quantity was expressed in one unit and the problem required the same quantity to be expressed as a measure with another unit—were simple, but other problems were not. Those students who cyphered to the rules of three were required to apply the different rules to solving problems associated with a large range of everyday activities. The minority of students who cyphered beyond the rules of three solved problems that might have been expected to arise in the world of commerce—some of these problems were associated with standard mercantile themes such as loss and gain, barter, brokerage, simple and compound interest, and discount, but others were associated with long-forgotten topics like tare and tret, alligation, fellowship, and false position.

One British government policy which influenced arithmetic curricula in the British colonies in North America related to a perceived need to achieve a favorable balance of trade with the colonies. From the British government's perspective, this implied that the government should import, from the North American colonies, produce and goods—tobacco, for example—that were in demand in Great Britain. In order to achieve the desired "balance of trade," the mother country exported items to the colonies that would not be missed at home. School textbooks written by British authors were in this category (Barber, 1976; Raven, 1997), and almost all of the arithmetic textbooks used in North American schools during the period 1607–1775 were written by British authors. In most cases, though not all, the books were printed in Great Britain before they were shipped to the colonies. Naturally, these textbooks were replete with examples involving sterling currency, and because of the flood of British arithmetic textbooks into North America it was not surprising that matters associated with British sterling currency occupied much space in cyphering books prepared by North American students.

Research Question 2: To What Extent Did Policies and Practices with Respect to Weights and Measures Influence School Arithmetic Curricula in the United States of America Before 1790?

Many of the comments made with respect to the first research question are also germane with respect to this second question. Sections relating to British weights and measures occupied much space in British arithmetic textbooks. Furthermore, since these British weights and measures were used in the counting houses in commercial and mercantile centers like Boston, Charleston, New York City, Philadelphia, Providence (Rhode Island) and Salem (Massachusetts), it was not surprising that material on weights and measures can be found on many pages of most North American cyphering books (see Tables 2.2 and 5.1 in this book for summaries of pertinent data).

It would be wrong, though, to attribute this emphasis in cyphering books on weights and measures solely to government policies. School arithmetic in the North American colonies took place more or less independently of government policy. There were no government inspectors of schools, and no government requirement on what should be included in arithmetic curricula. Influences on curriculum were hidden in ethnomathematical forces within the community—in the local produce markets, in the customs houses, on the wharves, and in the offices of captains of business. Working on the wharves in Salem, Massachusetts, for example, were many young employees who, while at school, had learned to cope with the arithmetic associated with tare and tret, loss and gain, practice, and the rules of three. Salem merchants and traders expected the town's local schools to provide youngsters with a thorough training in all the various forms of arithmetic associated with weights and measures. Those preparing cyphering books knew that they would be much more employable if they could show cyphering books in which there was a strong emphasis on sterling currency and on weights and measures (Ellerton & Clements, 2012; Walsh, 1801).

It is not an exaggeration to say that money, weights and measures, and standard applications of money and weights and measures constituted the core arithmetic content of the day. Only a small proportion of the students who prepared cyphering books included entries relating to "higher-order" topics like arithmetical and geometrical progressions, permutations and combinations, involution and evolution, mensuration, and logarithms.

Persons employed in mercantile activities, or employed as clerks were required to have an intimate knowledge of weights and measures, and those teaching school arithmetic were expected to take account of that unofficial expectation. In the eighteenth century there were no government policies by which arithmetic curricula in schools were defined, and so it would be wrong to argue that the schools' arithmetic curricula were directly affected by government policy. Any effects were indirect.

Research Question 3: To What Extent Did the Place of Decimal Fractions in U.S. School Arithmetic Curricula Change as a Result of the Introduction of a Decimalized Scheme of Federal Money Around 1790?

A full answer to this third research question can only be obtained by reading all of the chapters in this book. There can be no doubt that in the early 1780s Thomas Jefferson was convinced that the introduction of a decimalized form of Federal money would quickly and fundamentally change the teaching and learning of arithmetic in U.S. schools.

The evidence and analyses presented in this book suggest that, ultimately, Congress's decision to adopt Jefferson's and Madison's decimalized Federal money system did lead to far-reaching changes in intended and implemented arithmetic curricula in North America's schools. But, the wheels of curricular change turn slowly. Although most post-1792 North American cyphering books included entries on "Federal Money," for at least three decades the number of pages dedicated to tasks associated with Federal money was less than the number of pages dedicated to sterling currency tasks. The reason for the slow transformation was that the various state forms of sterling currency were designated "legal currency," and local traders often preferred to use the same kind of money transactions that they had been using for years (Goodwin, 2003). They were familiar with the arithmetic associated with "legal currency" and, in any case, they had worked out schemes by which they could profit from purchases and trades carried out with that form of currency. That was especially true in those cases when goods were purchased with one state's legal currency and sold with another's.

One of the fascinating findings of the present study was that even around 1860 the topic "decimal fractions" was studied as a topic in its own right by only about one-half of the students who prepared cyphering books. Teachers recognized that societal expectations meant they were more or less compelled to attempt to make their students familiar with calculations involving Federal money, but that did not mean that the teachers felt a need to teach decimal fractions as a topic. This was one of the things that Thomas Jefferson, in his willingness to predict what the effects of the introduction of decimal currency on ordinary people would be, got wrong. The introduction of Federal money did not result in most U.S. adults and school children quickly learning to apply decimal fractions concepts when performing tasks like finding the volume of wood in a pile whose length, width, and height were known.

Nevertheless, there can be no doubt that whereas before 1792 only a small proportion of North American cyphering books included entries on decimal fractions, after 1792, the proportion grew steadily (see Table 6.1). In Great Britain, where no form of decimal currency was introduced, the proportion of students who included pages on decimal fractions in their cyphering books decreased over the period 1793 through 1888.

Research Question 4: To What Extent Did Policies and Practices with Respect to Weights and Measures Play a Part in School Arithmetic in the United States of America Between 1791 and 1860?

In order to answer this question we studied entries in Tables 2.2, 5.1, 5.2, and 6.1. Entries in Tables 2.2, 5.1, and 5.2 showed the percentage of pages in North American cyphering books devoted entirely to weights and measures for the non-overlapping periods 1742–1791, 1792–1800, and 1801–1810, respectively.

During the period 1742–1791 the percentage of the pages in the cyphering books that were analyzed was 15%, and during the periods 1792–1800 and 1801–1810 the percentages were 19% and 21% respectively. So, there was a small increase in the emphasis on weights and measures over the period 1742–1810.

More qualitative, page-by-page analyses of cyphering-book data for the periods 1742–1791, 1792–1800, and 1801–1810 indicated that there was very little change in the emphases with respect to weights and measures. So far as weights and measures were concerned, the various sub-topics remained extremely important, and the same kinds of tasks and problems which were found in cyphering books in the period before 1792 were also found after 1792. Occasionally one would find an attempt to solve a problem using decimal methods, but generally speaking the old methods continued to be used. Even in textbooks, authors used non-decimal methods in the first whole-number part of the book. In the later sections on "vulgar arithmetic" and "decimal arithmetic," they made use of vulgar fractions and decimal fractions, respectively, but cyphering-book evidence shows that, often, teacher-implemented arithmetic curricula did not go as far as vulgar fractions and decimal fractions—for those topics appeared quite late in the standard *abbaco* sequence.

Thomas Jefferson wanted to introduce a Federal decimalized, coordinated system of weights and measures, and in the early 1790s he almost succeeded in achieving that. Undoubtedly, if he had succeeded, implemented arithmetic curricula in U.S. schools would have changed much more than they did.

Research Question 5: What are the Implications of the Findings of This Study for Mathematics Curriculum Theory?

Intended, Implemented and Attained Curricula

The modern distinction between intended, implemented and attained curricula (Westbury, 1980) offered a powerful way of thinking about the issues studied in the investigation described in this book. Its first and most obvious advantage for this study was that it permitted a sharp distinction between the kinds of arithmetic curricula that could be gleaned from studying school arithmetic textbooks, which were regarded as providing evidence with respect to the author-intended curricula, and the implemented curricula, which could be inferred by studying handwritten cyphering books prepared by students.

Our identification of the curricula found in textbooks with author-intended curricula needs to be interpreted. Certainly, our analysis of the topics included in arithmetic textbooks intended for schools in Great Britain and the United States throughout the seventeenth, eighteenth, and nineteenth centuries left us in little doubt that there was an underlying force at work which selected and ordered topics. Elsewhere we have called this the *abbaco*, or cyphering, tradition (Ellerton & Clements, 2012, 2014).

Post-dame-school arithmetic began with numeration and notation, and then would follow the four operations on whole numbers (addition, subtraction, multiplication, and division), compound operations (applying the four operations to money, and to weights and measures), reduction, practice and the rules of three. That was the elementary part of the *abbaco* curriculum, and those topics would appear, almost always in the order stated— although, perhaps, the positioning of "practice" might vary. Then would follow money topics—such as loss and gain, tare and tret, simple and compound interest, discount, barter, equations of payment, fellowship, and false position—treated from whole-number approaches only. The topics may not always have been presented in *exactly* that order, but differences were never great. Methods for solving the problems within these topics using whole numbers only were presented. That was the case even with questions on simple and compound interest, for which the concept of percentage was used—although that concept had never formally been studied by students—and rates of interest could be fractional (like, for example, $5\frac{3}{4}$ percent). At that stage, a few higher-order topics such as alligation, arithmetical and geometrical progressions might appear. Then would come a new section labeled "Vulgar Arithmetic," and all the topics covered under "Whole Number Arithmetic" would be revisited, with questions being framed so that vulgar fractions would be involved. And then would come a second revisiting, with decimal fractions being involved.

Anyone who studied the textbooks only would get the impression that the author-intended curricula included vulgar fractions and decimal fractions. But, analyses of cyphering books showed that often neither vulgar fractions nor decimal fractions were studied by students. That finding raised the question whether textbook authors intended that many students—perhaps a majority—would not go beyond the whole-number-arithmetic stage. We examined the authors' prefaces to finds hints with respect to that issue but did not find answers. Probably the authors knew that not all students would go beyond whole-number arithmetic, but they included the extra sections on vulgar fraction arithmetic and decimal fraction arithmetic for the more advanced students. The authors would have known that any school arithmetic which did not attend to vulgar fractions or decimal fractions would have been regarded as deficient by the educated elite.

When one studies a large number of students' cyphering books one cannot escape an initial feeling of authenticity. One cannot help but think that this was what the students *really* did, that this testified unambiguously to teacher-implemented arithmetic curricula. But, after studying many cyphering books, we began to question whether that initial very strong feeling of authenticity was always warranted. Often, there was evidence of extensive copying— which could have occurred by students copying from textbooks or from older, "parent," cyphering books. Often there were serious misspellings, suggesting that teachers dictated notes to students. But, evidence of copying alone was not sufficient to raise doubts about the legitimacy of applying the term "implemented curriculum," for that could have been the accepted way that cyphering books were prepared. The more difficult question arose when it appeared to be the case that the writing on many of the pages of a cyphering book was not done by the students whose names appeared on the covers of the cyphering book (and even on the insides of books where assertions that "This is my book" were often written). The beautiful cyphering book prepared between 1776 and 1781 and attributed to sisters Martha and Elisabeth Ryan, of North Carolina, would appear to be a case in point—it seems that often someone else solved the exercises, prepared the calligraphic headings, and wrote the notes (see Ellerton and Clements, 2014, Chapter 4, for a full discussion of the Ryan sisters' cyphering book). We reached the conclusion that one should not always assume that what appeared on the pages of a cyphering book represented what the student whose name was on the cover actually did in school arithmetic.

Despite the above-mentioned difficulties with respect to the concepts of intended and implemented curricula, we still concluded that the distinction between the two was a powerful one for this study. Textbooks were prepared by authors who were thinking of what was needed in schools, and cyphering books were prepared in schools by students, probably with variable amounts of assistance from teachers. So far as the attained curriculum was concerned, there are no examination or test data, or interview data, available from the eighteenth and early nineteenth centuries, and so analysis of students' cyphering books offers the best chance of finding out how much students actually learned. But, given the likelihood that many entries were copied, or were the result of dictation, or were even made by teachers, one cannot come close to knowing how well the students learned what appeared in their cyphering books. That said, when we examined individual cyphering books we usually got a strong sense about how well the students who prepared the books had understood the arithmetic in those books. Issues associated with attained curricula could fruitfully be studied by future researchers.

The Lag-Time Theoretical Lens

Scholars who investigate the history of school mathematics have rarely given much thought to the passage from the conception of decimal fractions to the stage where decimals fractions were regarded as a normal part of school mathematics curricula. From our perspective, however, that set of events has been worthy of special study, for decimal fractions have emerged as a core component of school mathematics. With the advent of logarithms, of decimal coinage, of the metric system, and of electronic calculators, decimal fractions are clearly of enormous practical importance, for almost everyone.

This study has revealed just how precarious the passage was from the theoretical development of decimal fractions—which we assume to have occurred toward the end of the sixteenth century—to a "decimals-for-all" state of affairs, when decimal fractions assumed an important place in most schools' arithmetic curricula. Clearly, lag time has varied enormously across the world. The analysis of British and North American cyphering books in Chapter 6 of this present work made clear that whereas decimal fractions slowly, yet steadily, became part of school arithmetic curricula in North American schools, that same kind of change did not occur in Great Britain. But even in the United States of America it would be wrong to say that by the middle of the nineteenth century decimal fractions were being studied by all children who prepared cyphering books.

As it has been presented in this book, the lag-time theoretical position emphasizes that lag time should vary depending on political, social and economic forces. Our conjecture, that the introduction of a decimalized form of Federal money into the United States late in the eighteenth century hastened the time when decimal fractions would enter the implemented curricula of most schools in the nation, has been supported—though not conclusively proven—by the analysis presented in this book. As the first half of the nineteenth century unfolded, a greater proportion of students in American schools than in British schools prepared pages in their cyphering books which dealt with decimal fractions. One wonders what would have happened, so far as the inclusion of decimal fractions into implemented curricula was concerned, if, in the early 1790s, Congress had brought into existence a national, decimalized system of weights and measures.

The fact that the introduction of decimal currency into North America did not result in teachers of arithmetic all across the nation immediately including decimal fractions in their implemented curricula provides testimony to the force of social tradition, and of curriculum inertial forces associated with teachers not being willing to take on new ideas unless they felt pressured to do so. With the introduction of a Federal decimalized money system some teachers felt compelled to include decimal fractions in their implemented arithmetic curricula, but many did not. The teachers were prepared to help their students learn to calculate with Federal money, but there was no need to go beyond this. For these teachers, decimal currency and decimal fractions were not as closely related as one might have expected them to be.

It will be the task of future historians to identify when decimal fractions became part of the implemented curricula for most learners in various nations. France, which introduced both a decimal currency and a decimalized system of weights and measures in the 1790s, should provide an especially interesting case study. Russia, which introduced a decimal form of currency early in the seventeenth century, would offer another interesting case study. And, one cannot help but wonder whether the introduction of decimalized forms of currency and weights and measures had any effect on when implemented arithmetic curricula in Continental Western European nations included decimal fractions. Two notable historians of mathematics education, David Eugene Smith and Louis Karpinski, said of the situation in Germany, "a traveler in Germany in the year 1700 would probably have heard or seen nothing of decimal fractions, although these were perfected a century before that date" (Smith & Karpinski, 1911, p. 45). Although they did not use the expression "lag time," Smith and Karpinski commented that the gap between when "the élite of mathematicians" knew some mathematical principle and when that same principle was known by "merchants and the common people" was likely to be a "long time" (p. 45), and that observation has been supported by the analyses in this present work.

The lag-time theoretical lens has generated interesting reflections and pushed the researchers in this study along pathways that, without that lens, they would not have taken. Thus, for example, the comparison of the uptake of decimal fractions in the United States and Great Britain, as presented in Chapter 6, would not have occurred if it had not been for the stimulus to carry out such an investigation provided by the lag-time theoretical lens.

Lag-time research in mathematics education could fruitfully be carried out with respect to many content areas. For instance, right across the world the uptake of graphs, on Cartesian planes, in school algebra was slow, and one wonders why.

Limitations of the Investigation

One of the most difficult aspects in the conduct of this research was that, as far as we could determine, nobody else had ever carried out research which resembled the research we were doing. There were three reasons for that:

1. No other researcher had ever had access to the same number of cyphering books that we held (in the Ellerton-Clements collection). The importance of that accessibility factor was magnified by the fact that our collection included 102 British cyphering-book units—in addition to the 370 North American cyphering-book units.

2. In addition to the cyphering book collection we also had ready and daily access to a large collection of arithmetic textbooks used in North America and in Great Britain in the eighteenth and nineteenth centuries. This access meant that whenever we felt the need to do so we could check author-intended curricula against teacher implemented curricula.

3. It seemed to us that other historians had not identified the history of the uptake of decimal fractions in North American and British schools as an important area needing to be studied.

At first, the conditions we have described in these three points generated, in us, the feeling that we were engaged in pioneering research. But those feelings quickly gave way to a feeling of intellectual loneliness. The issues we were investigating have rarely if ever been the focus of attention in books, or in journal articles, or in dissertations. We have never heard them discussed at conferences. As far as we know, no other mathematics educator has used the lag-time theoretical lens in a research investigation.

We regard our idiosyncratic approach to the research as a strength, but others could see it as a limitation. We recognize that our findings will need to be checked by later historians—who might adopt a more conventional approach to investigating the history of the uptake of decimal fractions in schools across the world.

As already indicated, we had at our disposal, throughout the research a very large collection of North American and British cyphering books. Notwithstanding, it would have been beneficial to locate and access a larger set of other primary sources (e.g., novels, private letters by students and teachers, and statements in school records). The identification consultation, and analysis of such "other" primary sources can be the work of later researchers who take up the themes addressed in this book.

Recommendations for Further Research

Today, decimal fractions are studied in schools right across the world, and not many people would question the wisdom of such a state of affairs. It would be an interesting study to find out when, for various nations (like for example, France, Great Britain, Russia, Greece, Japan, India, China, etc.), decimal fractions first came to be generally accepted as an important part of school mathematics. Reasons for why acceptance of decimal fractions into implemented curricula of some nations took longer to occur than in other nations would also be worthy of attention from researchers.

In this final chapter there has been no discussion of when and how vulgar fractions became part of the implemented curriculum in most schools in the United States of America. Given that vulgar fractions, when thought of as rational numbers, are vitally important in the structure of real numbers, a similar investigation to the one carried out in this book, only with the focus being on vulgar, rather than decimal fractions, could be illuminating. Many of the resources used in the present study (especially old textbooks, and cyphering books) would need to be consulted. Given that vulgar fractions have, traditionally been difficult for students, such an investigation might reveal the sources of some of the difficulties which students experience with vulgar fractions.

Other potentially useful investigations would compare the places of vulgar fractions and decimal fractions in author-intended and teacher-implemented school mathematics curricula. Also, from about 1850 onward, data with respect to how well students learned vulgar and decimal fractions might be available (in examination records held in different states), and hence it would be possible to investigate the *attained* curricula, using both quantitative and qualitative data sources. John Roach's (1971) research on written examinations in Great Britain would provide an excellent vantage point from which to begin contemplating such research.

Outside the areas of vulgar and decimal fractions there are many different themes within present-day mathematics curricula which might be usefully investigated from an intended/implemented/attained curriculum perspective. Topics which immediately spring to mind are: negative integers; the elementary geometry of polygons; Cartesian graphs (in algebra); the real number line; percentage; elementary trigonometry; limits (in the study of calculus); functions; and explicit and recursive sequences.

Whether or not researchers choose to take on, and perhaps modify, our lag-time theoretical lens will be up to them, but we would be especially interested in research which did use that theoretical framework.

This is the third book in which we have been primarily concerned with the history of school mathematics in North America in the seventeenth, eighteenth and nineteenth centuries. We look forward to a time when others will tackle some of the issues—both theoretical and practical—that we have had to face. Recent research on British cyphering books by John Denniss (2012), Jacqueline Stedall (2012), and Benjamin Wardhaugh (2012) has encouraged us to continue our research in the field, and we would hope that researchers, from a range of nations, will join us.

References

Barber, G. (1976). *Books from the old world and for the new: The British international trade in books in the eighteenth century.* Oxford, UK: Voltaire Foundation.

Boyd, J. P. (1961). *The papers of Thomas Jefferson, Volume 16, November 1789 to July 1790.* Princeton, NJ: Princeton University Press.

Cohen, P. C. (1982). *A calculating people: The spread of numeracy in early America.* Chicago, IL: University of Chicago Press.

Dantzig, T. (1954). *Number the language of science: A critical survey written for the cultured non-mathematician.* New York, NY: Macmillan.

Denniss, J. (2012). *Figuring it out: Children's arithmetical manuscripts 1680–1880.* Oxford, UK: Huxley Scientific Press.

Dilworth, T. (1773). *The schoolmasters assistant: Being a compendium of arithmetic, both practical and theoretical* (17th ed.). Philadelphia, PA: Joseph Cruikshank.

Ellerton, N. F., & Clements, M. A. (2012). *Rewriting the history of school mathematics in North America 1607–1861.* New York, NY: Springer.

Ellerton, N. F., & Clements, M. A (2014). *Abraham Lincoln's cyphering book and ten other extraordinary cyphering books.* New York, NY: Springer.

Goodwin, J. (2003). *Greenback: The almighty dollar and the invention of America.* New York, NY: Henry Holt and Company.

Jefferson, T. (1784). *Notes on the establishment of a money unit and of a coinage for the United States* (handwritten manuscript). Washington, DC: Library of Congress.

Linklater, A. (2003). *Measuring America: How the United States was shaped by the greatest land sale in history.* New York, NY: Plume

Michael, I. (1993). The textbook as a commodity: Walkingame's *The Tutor's Assistant. Paradigm, 12,* 2–10.

Raven, J. (1997). The export of books to colonial North America. *Publishing History, 42,* 11–49.

Roach, J. (1971). *Public examinations in England 1850–1900.* Cambridge, UK: Cambridge University Press.

Simons, L. G. (1936). Short stories in colonial geometry. *Osiris, 1,* 584–605.

Stedall, J. (2012). *The history of mathematics: A very short introduction.* Oxford, UK: Oxford University Press.

Walkingame, F. (1785). *The tutor's assistant: Being a compendium of arithmetic and a complete question book* (21st ed.). London, UK: J. Scratcherd & I. Whitaker.

Walsh, M. (1801). *A new system of mercantile arithmetic: Adapted to the commerce of the United States, in its domestic and foreign relations; with forms of accounts, and other writings usually occurring in trade.* Newburyport, MA: Edmund M. Blunt.

Wardhaugh, B. (2012). *Poor Robyn's prophecies: A curious almanac, and the everyday mathematics of Georgian Britain.* Oxford, UK: Oxford University Press.

Westbury, I. (1980). Change and stability in the curriculum: An overview of the questions. In H. G. Steiner (Ed.), *Comparative studies of mathematics curricula: Change and stability 1960–1980* (pp. 12–36). Bielefeld, Germany: Institut für Didaktik der Mathematik-Universität Bielefeld.

Author Biographies

M. A. (Ken) Clements's masters and doctoral degrees were from the University of Melbourne, and at various times in his career he has taught, full-time, in primary and secondary schools, for a total of 15 years. He has taught in six universities, located in three nations, and is currently professor in the Department of Mathematics at Illinois State University. He has served as a consultant and as a researcher in Australia, Brunei Darussalam, India, Malaysia, Papua New Guinea, South Africa, Thailand, the United Kingdom, the United States of America, and Vietnam. He served as co-editor of the three *International Handbooks of Mathematics Education*—published by Springer in 1996, 2003 and 2013—and, with Nerida Ellerton, co-authored a UNESCO book on mathematics education research. He has authored or edited 31 books and over 200 articles on mathematics education, and is honorary life member of both the Mathematical Association of Victoria and the Mathematics Education Research Group of Australasia.

Nerida F. Ellerton has been professor in the Department of Mathematics at Illinois State University since 2002. She holds two doctoral degrees—one in Physical Chemistry and the other in Mathematics Education. Between 1997 and 2002 she was Dean of Education at the University of Southern Queensland, Australia. She has taught in schools and at four universities, and has also served as consultant in numerous countries, including Australia, Bangladesh, Brunei Darussalam, China, Malaysia, the Philippines, Thailand, the United States of America, and Vietnam. She has written or edited 16 books and has had more than 150 articles published in refereed journals or edited collections. Between 1993 and 1997 she was editor of the *Mathematics Education Research Journal*, and since 2011 she has been Associate Educator of the *Journal for Research in Mathematics Education*. In 2012, and in 2013, Springer published the 223-page *Rewriting the History of School Mathematics in North America 1607–1861* and the 367-page *Abraham Lincoln's Cyphering Book and Ten other Extraordinary Cyphering Books,* respectively. She jointly authored both of those books with M. A. (Ken) Clements.

M. A. (Ken) Clements, & N. F. Ellerton, *Thomas Jefferson and his decimals 1775–1810: Neglected years in the history of U.S. school mathematics*, DOI 10.1007/978-3-319-02505-6,
© Springer International Publishing Switzerland 2015

Appendix A

Notes on Thomas Jefferson's (1784) "Some Thoughts on a Coinage"

The two-page document, "Some Thoughts on a Coinage," is printed in Volume 7 of Julian P. Boyd's (1953), *The Papers of Thomas Jefferson* (Vol. 7 March 1784 to February 1785), published by Princeton University Press. Jefferson's original handwritten pages were undated, but Boyd indicated that they were prepared around March 1784—that is to say, *before* Jefferson left for Paris, in July 1784, to serve as Minister Plenipotentiary to France. According to Boyd (1953), "Some Thoughts on a Coinage" was written entirely in Jefferson's hand, and was "erroneously placed with the rough draft and notes of Jefferson's report to the House of Representatives of a plan for establishing uniformity in currency, weights and measures" dated July 4, 1790 (p. 175). The document is in the Library of Congress, Washington, DC, Thomas Jefferson Papers, 233, 41972.

"Some Thoughts on a Coinage" needs to be distinguished from the better-known and much longer "Notes on Coinage," which was written between March and May 1784 (and is printed in Boyd, 1953, pages 175–185). With respect to "Some Thoughts on a Coinage," Boyd (1953) wrote:

> It is now known that Jefferson considered at this time a comprehensive plan for the decimalization of weights and measures as well as money. Document III ["Some Thoughts on a Coinage"], never before published, shows that Jefferson's "Some Thoughts on a Coinage" was in reality an outline of "Notes on Coinage." It was probably drawn up as early as March, or even February, 1784. At that time Jefferson must have intended to advocate the dollar as the money unit as well as a decimalized coinage, and once these points were established, to make a transition to a decimal reckoning in weights, measures, and perhaps time. But he must have concluded that the country was not ready for such a thoroughgoing innovation and that the latter parts of his program could be more readily accomplished after the coinage had been settled. In this sense, then, Jefferson's "Notes on Coinage" must be regarded merely as the preliminary expression of a plan to which he returned six years later, immediately on assuming office as Secretary of State. (pp. 155–156).

Here is how Boyd's (1953) reprinting of "Some Thoughts on a Coinage" appeared. Jefferson's misspellings and idiosyncratic use of upper-case first letters have been retained.

III. Some Thoughts on a Coinage

[ca. Mch 1784]

Some Thoughts on a Coinage and the Money Unit for the U.S.
1. The size of the Unit.
2. It's division.
3. It's accommodation to known coins.

The value of a fine silver in the Unit.

M. A. (Ken) Clements, & N. F. Ellerton, *Thomas Jefferson and his decimals 1775–1810: Neglected years in the history of U.S. school mathematics*, DOI 10.1007/978-3-319-02505-6,
© Springer International Publishing Switzerland 2015

The proportion between the value of gold and silver.
   The alloy of both 1. oz in the pound. This is Brit. standard of gold, and Fr[ench] Ecu of silver.

The Financier's plan.

A table of the value of every coin in Units.

Transition from money to weights.
   10 Units to the American pound. 3650 grs = 152 dwt. 2 grs. = 7 oz. 12 dwt. 2 grs

Transition from weights to measures.
Rain water weighing a pound, i.e. 10 Units, to be put in a cubic vessel and one side of that taken for the standard or Unit of measure.
   Note. By introducing pure water and pure silver, we check errors of calculation proceeding from heterogeneous mixtures with either.

Transition for measures to time.

I find new dollars of 1774, 80, 81 (qu. Mexico Pillar) weigh 18 dwt. 9 grs. = 441 grs. If of this there be but 365 grs. Pure silver, the alloy would be 2.1 oz. in the lb. instead of 19 dwt. The common Spanish alloy, which is 1 dwt. Worse than the Eng. Standard. Whereas if it is of 19 dwt. In the lb. troy, it will contain 406 grs. Pure silver. The Seville piece of eight weighing 17½ dwt. By Sir I. Newt's assay contained 387. grs. Pure silver. The Mexico peice of 8 [weighing] 17 dwt. $10\frac{5}{9}$ grs. (alloy 18 dwt. as the former) 385½. The Pillar peice of 8 [weighing] 17–9 (alloy 18 dwt.) 385¾. The old ecu of France or peice of 6, gold Tournois is exactly of the weight and fineness of the Seville peice of 8. The new ecu is by law 1. oz. alloy, but in fact only 19½ dwt. 19 dwt. 14½ grs. pure metal is 432¼ grs.

Dollars	Weight	In water	Loss
[a]1773	17—8½	15—15	1—17½
[a]1774	17—8	15—14	1—18
[a]1775	17—8½	15—15½	1—17
1776			
1777			
[a]1778	17—9½	15—16	1—17½
[a]1779	17—9½	15—16	1—17½
[a]1780	17—10	15—17½	1—16½
[a]1781	17—8½	15—15½	1—17
1782			

[a]These average 417. grs. weight in air 41.3 grs. loss in water. i.e. 1/10 or nearly 1/1000 [Jefferson meant 100/1000] or ten times the weight of water. Cassini makes a degree in a great mile contain

Miles   D
69   864 = 365,184 feet
Then a geographical mile will be 6086.4 feet.
        a Statute mile is 5280 f.
A pendulum vibrating seconds is by Sr. I. Newton 39.2 inches = 3.2666 &c. feet
Then a geographical mile of 6086.1 f. = 1863 second pendulums.
Divide the geometrical mile into 10. furlongs
        each furlong 10 chains
        each chain 10. paces
Then the American mile = 6086.4 f.   English = 5280 f.
        furlong   = 608.64 f.   = 660
        chain   = 60.864   = 66
        pace   = 6.0864   fathom = 6.

Russian mile	.750 of a geographical mile
English mile	.8675
Italian mile	1.
Scotch and Irish do.	1.5
old league of France	1.5
small league of do.	2.
great league of do.	3.
Polish mile	3.
German mile	4.
Swedish mile	5.
Danish mile	5.
Hungarian mile	6

A rod vibrating seconds is nearly 58½ inches.

Comments by Clements and Ellerton (2014)

"Some Thoughts on a Coinage" is a historically remarkable document that reveals Jefferson's struggles to achieve a scientific and mathematical overview of the situation with respect to coinage and weights and measures. Reading the document makes it obvious that before he went to Paris Jefferson was extremely well read with respect to the key issues, and that he thought a totally integrated system of weights and measures and money was what the newly-created United States of America should strive to achieve. It is also obvious that in "Some Thoughts on a Coinage" he is not merely reiterating what someone else has advised him. He had taken on the challenge of devising a coordinated system that fitted the demands of science and mathematics, yet at the same time would be convenient in day-to-day practices.

"Some Thoughts on a Coinage" should be read in conjunction with the slightly later, and better known "Notes on Coinage" (see Boyd, 1953. pp. 175–185, as well as Julian Boyd's (1953) editorial notes pp. 150–160 and pp. 185–188). It is our view that Jefferson probably taught the French thinkers on weights and measures more about the possibilities of a coordinated system of weights and measures than they taught him.

The principles of the new coinage had been established by Jefferson before he left for France, but it still needed to be passed through Congress. While he was in France, James Madison, Jefferson's friend who would become the fourth U.S. President (1809–1817), took on the role of leader with respect to need to move toward a decimalized system of coinage and weights and measures. On April 28, 1785, Madison wrote to James Monroe, another future U.S. President (1817–1825), expressing his concerns about currency, weights, and measures:

> I hear frequent complaints of the disorders of our coin, and the want of uniformity in the denominations of the States. Do not Congress think of a remedy for these evils? The regulation of weights and measures seem also to call for their attention. Every day will add to the difficulty of executing these works. If a mint be not established and a recoinage effected while the federal debts carry the money through the hands of Congress, I question much whether their limited powers will ever be able to render this branch of their prerogative effectual. With regard to the regulation of weights and measures, would it not be highly expedient, as well as honorable to the federal administration, to pursue the hint which has been suggested by ingenious and philosophical men, to wit: that the standard of measure should be first fixed by the length of a pendulum vibrating seconds at the Equator or any given latitude; and that the standard of weights should be a cubical piece of gold, or other homogeneous body, of dimensions fixed by the standard of measure? Such a scheme appears to be easily reducible to practice; and as it is founded on the division of time, which is the same at all times and in all places, and proceeds on other data which are equally so, it would not only secure a perpetual uniformity throughout the United States, but might lead to universal standards in these matters among nations. Next to the inconvenience of speaking different languages, is that of using different and arbitrary weights and measures.

(Quoted in Hunt, 1901, p. 142)

The Coinage Act of 1792 (Garrett & Guth, 2003; Goodwin, 2003) would introduce the following names for amounts or coins which would be decimal fractions of a dollar—cent, mill, and disme (dime). Those names, which Jefferson proposed before he went to France, clearly predated similar Latin prefixes which were part of the terminology that was incorporated in France's metric system.

References

Boyd, J. P. (Ed.). (1953). *The papers of Thomas Jefferson* (Vol. 7, March 1784 to February 1785). Princeton, NJ: Princeton University Press.

Garrett, J., & Guth, R. (2003). *100 greatest U.S. coins*. Atlanta, GA: H.E. Harris.

Goodwin, J. (2003). *Greenback: The almighty dollar and the invention of America*. New York, NY: Henry Holt.

Hunt, G. (Ed.). (1901). *The writings of James Madison: 1783–1787*. New York, NY: G. P. Putnam.

"III. Some Thoughts on a Coinage [ca. March 1784]." Founders Online, National Archives (http://founders.archives.gov/documents/Jefferson/01-07-02-0151-0004, ver. 2014-05-09). Source: *The papers of Thomas Jefferson, Vol. 7, 2 March 1784–25 February 1785*, Julian P. Boyd (Ed.), Princeton, NJ: Princeton University Press, 1953, pp. 173–175.

Appendix B

Comment on a Letter, Sent in March 1819, by Thomas Jefferson to Robert Patterson (1743–1824), Concerning Patterson's *A Treatise of Practical Arithmetic*

Robert Patterson was born in Ireland in 1743, and after early military service immigrated to North America in 1768, where he took up teaching. He was mathematically precocious, and in 1774 was made head of Wilmington Academy, in Delaware. He became Professor of Mathematics at the University of Pennsylvania in 1782, and remained in that position until 1813, holding the title Professor of Natural Philosophy and Mathematics between 1810 and 1813. He served as Secretary, Vice-President and President of the American Philosophical Society, and was Director of the U.S. Mint. He knew Thomas Jefferson well.

In 1818 and 1819 Patterson authored a two-volume school arithmetic (*A Treatise of Practical Arithmetic Intended for the Use of Schools*) which was published by R. Patterson and Lambdin. The volumes, which occupied 390 pages, did not sell well, and no additional editions appeared. A full *abbaco* curriculum was presented in the *Treatise* but, from our perspective, the material was written in a way, and at a level, more suited to university students than to most school students. More advanced *abbaco* topics such as involution and evolution, arithmetical and geometrical progressions, logarithms, annuities, permutations and combinations, trigonometry, mensuration of superficies and solids, number theory, and gauging were dealt with, as were a number of "select problems in natural philosophy—such as "oscillation of pendulums," and "problems respecting the mechanic powers" (Patterson, 1819).

Patterson sent copies of the two parts of *A Treatise of Practical Arithmetic* to Thomas Jefferson, and on March 14, 1819 Jefferson replied, thanking Patterson for the copies. In December 2012 Jefferson's letter was sold by Cowan's Auctions (see http://www.cowanauctions.com/auctions/item.aspx?ItemId=120004).

In his letter of reply, Jefferson stated that Patterson's arithmetic was consistent with his view that the State of Virginia should develop a three-level system of education. At the first, and lowest, level, "primary" schools would teach reading, writing and arithmetic and be available to children of all citizens. At the second, intermediate level would be district colleges, which would offer to selected students from the primary schools in the district more advanced teaching in the classics, the French language, and more advanced arithmetic. And, at the third and highest level, there would be a state university, which would deal with higher sciences. Jefferson commented in his letter that Patterson's book would be useful for the first two levels of his scheme, although a more sophisticated treatment in some of the topics might be needed so that the text would reach as high as standard as he hoped for in the second-level colleges.

As we wrote in Chapter 3 of this book, Jefferson overestimated the level of arithmetic suitable for most school students. Although he thought he knew how students would solve problems, often he did not. But, in relation to a system of decimal coinage, he got the "big picture" right, and that was what mattered most.

M. A. (Ken) Clements, & N. F. Ellerton, *Thomas Jefferson and his decimals 1775–1810: Neglected years in the history of U.S. school mathematics*, DOI 10.1007/978-3-319-02505-6,
© Springer International Publishing Switzerland 2015

Figure B1 shows a typical page from Patterson's (1819) *A Treatise of Practical Arithmetic Intended for the Use of Schools.* Note the heavy use of algebra at a time when only a very small minority of school students studied algebra.

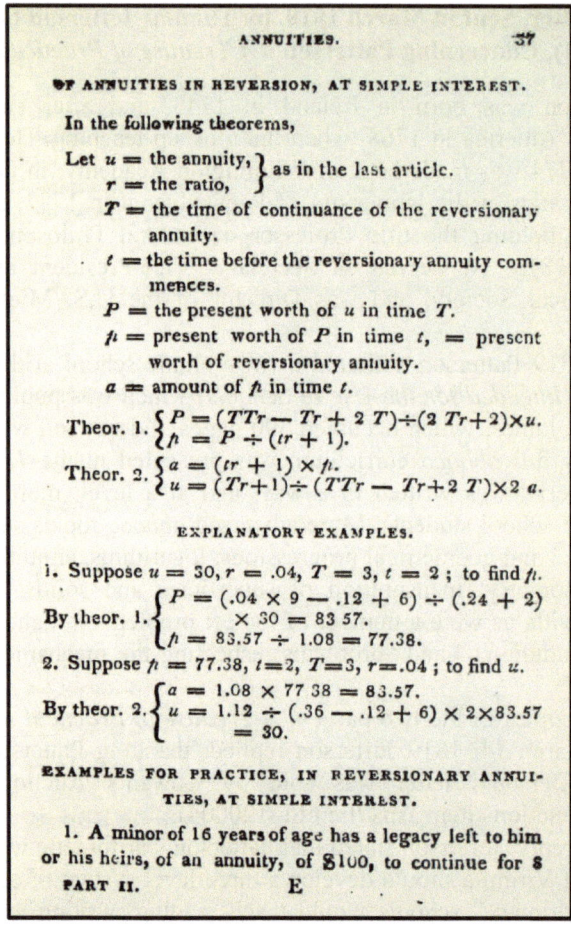

Figure B1. Page 37 of Part 2 of Robert Patterson's (1819) *A treatise of practical arithmetic intended for the use of schools*

References

Patterson, R. (1819). *A treatise of practical arithmetic intended for the use of schools.* (2 parts). Pittsburgh, PA: R. Patterson and Lambdin.

Thomas Jefferson to Robert Patterson. March 14, 1819 (sold December 7, 2012, see http://www.cowanauctions.com/auctions/item.aspx?ItemId=120004).

Combined References List

[*Note: Cyphering books in the Ellerton-Clements and in other cyphering-book collections will not be included in this Combined Reference List.*]

Adams, D. (1801). *The scholar's arithmetic—Or Federal accountant.* Leominster, MA: Adams and Wilder.

A Lady of Massachusetts. (1798). *The boarding school, or lessons of a preceptress to her pupils: Consisting of information, instruction, and advice, calculated to improve the manners, and form the character of young ladies.* Boston, MA: I. Thomas and E. T. Andrews.

Albree, J. (2002). Nicolas Pike's Arithmetic (1788) as the American *Liber Abbaci.* In D. J. Curtin, D. E. Kullman, & D. E. Otero (Eds.), *Proceedings of the Ninth Midwest History of Mathematics Conference.*(pp. 53–71). Miami, FL: Miami University.

Alder, K. (2002). *The measure of all things: The seven-year odyssey and hidden error that transformed the world.* New York, NY: The Free Press.

Allen, L. (2009). *The encyclopedia of money* (2nd ed.). Santa Barbara, CA: ABC-CLIO.

Barber, G. (1976). *Books from the old world and for the new: The British international trade in books in the eighteenth century.* Oxford, UK: Voltaire Foundation.

Barnard, H. C. (2008). *Education and the French Revolution.* Cambridge, UK: Cambridge University Press.

Bartlett, J. R. (1933). *Letter of instructions to the captain and the supercargo of the brig "Agenoria," engaged in a trading voyage to Africa.* Philadelphia, PA: Howard Greene and Arnold Talbot.

Beck, L. W. (1966). Introduction. In L. W. Beck (Ed.), *Eighteenth-century philosophy* (pp. 1–11). New York, NY: The Free Press.

Bellman, R., & Danskin, J. M. (1954). *A survey of the mathematical theory of time-lag, retarded control, and hereditary processes.* The RAND Corporation, Report R-256.

Besomi, D. (1998) Harrod, and the "time-lag" theories of the cycle. In G. Rampa, S. Stella, & A. P. Thirlwall (Eds.), *Economic dynamics, trade and growth: Essays on Harrodian themes* (pp. 107–143). Basingstoke, UK: Macmillan.

Bochner, S. (1966). *The role of mathematics in the rise of science.* Princeton, NJ: Princeton University Press.

Bordley, John Beale (1789). *On monies, coins, weights and measures.* Philadelphia, PA: Daniel Humphreys.

Bowditch, N. (1797). *Journal of a voyage from Salem to Manila in the ship Astrea, E. Prince, Master, in the years 1796 and 1797* (Handwritten manuscript held in the Bowditch Collection). Boston, MA: Boston Public Library.

Bowen, C. D. (2007). *Miracle at Philadelphia: The story of the constitutional convention, May to September 1787.* New York, NY: Little Brown & Co.

Boyd, J. P. (Ed.) (1950a). *The papers of Thomas Jefferson* (Vol. 1, 1760–1776). Princeton, NJ: Princeton University Press.

Boyd, J. P. (Ed.). (1950b). *The papers of Thomas Jefferson* (Vol. 2, January 1777 to June 1779). Princeton, NJ: Princeton University Press.

M. A. (Ken) Clements, & N. F. Ellerton, *Thomas Jefferson and his decimals 1775–1810: Neglected years in the history of U.S. school mathematics,* DOI 10.1007/978-3-319-02505-6,
© Springer International Publishing Switzerland 2015

Boyd, J. P. (Ed.). (1953). *The papers of Thomas Jefferson* (Vol. 7, March 1784 to February 1785). Princeton, NJ: Princeton University Press.

Boyd, J. P. (1961a). Report on weights and measures: Editorial note. In J. Boyd (Ed.), *The papers of Thomas Jefferson 16, November 1789 to July 1790* (pp. 602–617). Princeton, NJ: Princeton University Press.

Boyd, J. P. (Ed.), (1961b). *The papers of Thomas Jefferson* (Vol. 16, November 1789 to July 1790). Princeton, NJ: Princeton University Press.

Boyer, C. B. (1991). *A history of mathematics* (2nd ed.). New York, NY: Wiley.

Bremner, R. H. (1971).*Children and youth in America: A documentary history, Vol. 1, 1600–1865.* London, UK: Oxford University Press.

Briggs, H. (1617). *Logarithmorum chilias prima.* London, UK: Author.

Brock, L. V. (1975). *The currency of the American colonies, 1700–1764: A study in colonial finance and imperial relations.* New York, NY: Arno Press.

Buer, M. C. (1926). *Wealth and population in the early days of the industrial revolution,* London, UK: Routledge.

Burton, J. (1794). *Lectures on female education and manners.* New York, NY: Samuel Campbell.

Burton, W. (1833). *The district school as it was, by one who went to it.* Boston, MA: Carter, Hendee and Company.

Cajori, F. (1907). *A history of elementary mathematics with hints on methods of teaching.* New York, NY: Macmillan.

Cobb, L. (1835). *Cobb's ciphering book, No. 1, containing all the sums and questions for theoretical and practical exercises in Cobb's explanatory arithmetick No. 1.* Elmira, NY; Philadelphia, PA: Birdsall & Co; James Kay Jn & Brother. (stereotyped by J. S. Redfield).

Cocker, E. (1677). *Cocker's arithmetick: Being a plain and familiar method suitable to the meanest capacity for the full understanding of that incomparable art, as it is now taught by the ablest school-masters in city and country.* London, UK: John Hawkins.

Cocker, E. (1685). *Cocker's decimal arithmetick, ...* London, UK: J. Richardson.

Cocker, E. (1719). *Cocker's arithmetic: Being a plain and familiar method suitable to the meanest capacity ...* London, UK: H. Tracey.

Cocker, E. (1738). *Cocker's arithmetick.* London, UK: A. Bettesworth & C, Hitch.

Cogley, R. W. (1999). *John Eliot's mission to the Indians before King Philip's War.* Cambridge, MA: Harvard University Press.

Cogliano, F. D. (2008). *Thomas Jefferson: Reputation and legacy.* Charlottesville, VA: University of Virginia Press.

Cohen, P. C. (1982). *A calculating people: The spread of numeracy in early America.* Chicago, IL: University of Chicago Press.

Colburn, W (1821). *An arithmetic on the plan of Pestalozzi, with some improvements.* Boston, MA: Cummings and Hilliard.

Conant, J. B. (1962). *Thomas Jefferson and the development of American public education.* Berkeley, CA: University of California Press.

Cremin, L. A. (1970). *American education: The colonial experience 1607–1783.* New York, NY: Harper & Row.

Cubberley, E. P. (1962). *Public education in the United States.* Boston, MA: Houghton Mifflin Co.

Cutter, W. R. (Ed.). (1914). *New England families genealogical and memorial: A record of achievements of her people in the making of the Commonwealth and the founding of a nation.* New York, NY: Lewis Historical Publishing Company.

Daboll, N. (1800). *Daboll's schoolmaster's assistant: Being a plain, practical system of arithmetic; adapted to the United States.* New London, CT: Samuel Green.

Dantzig, T. (1954). *Number the language of science: A critical survey written for the cultured non-mathematician.* New York, NY: Macmillan.

De Morgan, A. (1847). *Arithmetical books, from the invention of printing to the present time.* London, UK: Taylor and Walton.

Denniss, J. (2012). *Figuring it out: Children's arithmetical manuscripts 1680–1880.* Oxford, UK: Huxley Scientific Press.

Devlin, K. (2011). *The man of numbers: Fibonacci's arithmetic revolution.* New York, NY: Walker & Company.

Dilworth, T. (1762). *The schoolmaster's assistant. Being a compendium of arithmetic, both practical and theoretical* (11th ed.). London, UK: Henry Kent.

Dilworth, T. (1773). *The schoolmasters assistant: Being a compendium of arithmetic, both practical and theoretical* (17th ed.). Philadelphia, PA: Joseph Cruikshank.

Dilworth, T. (1784). *The young book-keeper's assistant: Shewing him, in the most plain and easy manner, the Italian way of stating debtor and creditor* (9th ed.). London, UK: Richard and Henry Causton.

Dilworth, T. (1790). *The schoolmaster's assistant: Being a compendium of arithmetic both practical and theoretical.* Philadelphia, PA: Joseph Crukshank.

Dilworth, T. (1797). *The schoolmaster's assistant: Being a complete system of practical arithmetic.* Philadelphia, PA: Joseph Crukshank.

Dilworth, T. (1802). *The schoolmaster's assistant: Being a compendium of arithmetic both practical and theoretical.* Philadelphia, PA: Joseph Crukshank.

Dollarhide, W. (2001). *The census book: A genealogist's guide to Federal census facts, schedules and indexes.* North Salt Lake, UT: HeritageQuest.

Douglas, P. H. (1921). *American apprenticeships and industrial education.* PhD dissertation, Columbia University, New York.

Earle, A. M. (1899). *Child-life in colonial days.* New York, NY: The Macmillan Company.

Edmonds, M. J. (1991). *Samplers and sampler makers: An American schoolgirl art, 1700–1850.* New York, NY: Rizzoh.

Ellerton, N. F., Aguirre Holguín, V., & Clements, M. A (2014). He would be good: Abraham Lincoln's early mathematics, 1819–1826. In N. F. Ellerton & M. A. Clements (Eds.), *Abraham Lincoln's cyphering book and ten other extraordinary cyphering books* (pp. 123–186). New York, NY: Springer.

Ellerton, N. F., & Clements, M. A. (2008). An opportunity lost in the history of school mathematics: Noah Webster and Nicolas Pike. In O. Figueras, J. L. Cortina, S. Alatorrw & A. Mepúlveda (Eds.), *Proceedings of the Joint Meeting of PME 32 and PME-NA XXX* (Vol. 1, pp. 447–454). Morelia, Mexico: Cinvestav-UMSWH.

Ellerton, N. F., & Clements, M. A. (2012). *Rewriting the history of school mathematics in North America 1607–1861.* New York, NY: Springer.

Ellerton, N. F., & Clements, M. A. (2013, September). *The mathematics of decimal fractions and their introduction into British and North America schools.* Paper presented to the Third International Conference on the History of Mathematics Education held at Uppsala University, Sweden.

Ellerton, N. F., & Clements, M. A. (2014). *Abraham Lincoln's cyphering book and ten other extraordinary cyphering books.* New York, NY: Springer.

Ellis, J. R. (2000). *Founding brothers*. New York, NY: Random House.

Fanning, D. F., & Newman, E. P. (2011, July 24). More on the *American accomptant* and the first printed dollar sign. The *E-Sylum, 14*. Numosmatic Biblomania Society.

Farrington, F. (1910). *French secondary schools: An account of the origin, development and present organization of secondary education in France*. New York, NY: Longmans, Green and Co.

Fauvel, J. (1999, April 15). *Thomas Jefferson and mathematics*. Lecture given at the University of Virginia.

Fenzi, G. (1905). *The rubles of Peter the Great*. Moscow, Russia: Open Library.

Forsyth, M. I. (1913), The burning of Kingston, New York. *The Journal of American History, 7*(3), 1137–1145.

Frost, J. (1846). *Lives of American merchants*. New York, NY: Saxon & Miles.

Garrett, J., & Guth, R. (2003). *100 greatest U.S. Coins*. Atlanta, GA: H.E. Harris & Co.

George Washington to Nicolas Pike, June 20, *1788* (in Electronic Text Center, University of Virginia Library). Retrieved December 7, 2006, from http:etext.virginia.edu/etcbin/toccer new2?id=WasFi30.xml&ima

Gies, J., & Gies, F. (1969). *Leonard of Pisa and the new mathematics of the Middle Ages*. New York, NY: Thomas Y. Crowell.

Glaisher, J. W. L. (1873). On the introduction of the decimal point into arithmetic. *Report of the Meeting of the British Association for the Advancement of Science, 43*, 13–17.

Gomez, M. A. (1998). *Exchanging our country marks: The transformation of African Identities in the colonial and Antebellum South*. Chapel Hill, NC: University of North Carolina.

Good, H. G. (1960). *A history of Western education*. New York, NY: Macmillan.

Goodrich, S. (1857). *A pictorial history of the United States*. Philadelphia, PA: E. H. Butler & Co.

Goodwin, J. (2003). *Greenback: The almighty dollar and the invention of America*. New York, NY: Henry Holt and Company.

Greene, J. P., & Jellison. R. M. (1961). The Currency Act of 1764 in imperial-colonial relations, 1764–1776. *The William and Mary Quarterly, 18*(4), 485–518.

Greenwood, I. (1729). *Arithmetick, vulgar and decimal, with the application thereof to a variety of cases in trade and commerce*. Boston, MA: S. Kneeland & T. Green.

Grendler, P. F. (1989). *Schooling in Renaissance Italy literacy and learning, 1300–1600*. Baltimore, MD: Johns Hopkins University Press.

Grew, T. (1758). *The description and use of the globes, celestial and terrestrial; with variety of examples for the learner's exercise: ...* Germantown, PA: Christopher Sower.

Hadden, R. W. (1994). *On the shoulders of merchants: Exchange and mathematical conception of nature*. New York, NY: SUNY Press.

Harper, E. P. (2010). Dame schools. In T. Hunt, T. Lasley, & C. D. Raisch (Eds.), *Encyclopedia of educational reform and dissent* (pp. 259–260). Thousand Oaks, CA: Sage Publications.

Hepburn, A. B. (1915). *A history of currency in the United States with a brief description of the currency systems of all commercial nations*. New York, NY: The Macmillan Company.

Hill, J. (1772). *Arithmetick, both in the theory and practice, made plain and easy in the common and useful rules ...* London, UK: W. Strahan.

Honeywell, R. J. (1931). *The educational work of Thomas Jefferson.* Cambridge, MA: Harvard University Press.

Howson, G., & Rogers, L. (2014). Mathematics education in the United Kingdom. In A. Karp & G. Schubring (Eds.), *Handbook on the history of mathematics education* (pp. 257–282). New York, NY: Springer.

Høyrup, J. (2008). The tortuous ways toward a new understanding of algebra in the Italian *Abbacus* school (14th –16th centuries). In O. Figueras, J. L Cortina, A. Alatorre, T. Rojano & S. Sepulveda (Eds.), *Proceedings of the joint meeting of PME 32 and PME-NA XXX* (Vol. 1, pp. 1–20), Morelia, Mexico.

Høyrup, J. (2014). Mathematics education in the European Middle Ages. In A. Karp & G. Schubring (Eds.), *Handbook on the history of mathematics education* (pp. 109–124). New York, NY: Springer.

Humphreys, D. (1991). *Life of General Washington.* Athens, GA: University of Georgia Press.

Hunt, G. (Ed.). (1901). *The writings of James Madison: 1783–1787.* New York, NY: G. P. Putnam's Sons.

Jackson, L. L. (1906). *The educational significance of sixteenth century arithmetic from the point of view of the present time.* New York, NY: Teachers College Columbia University.

Jefferson, T. (1784a). *Notes.* In W. Peden (Ed.), *Notes on the State of Virginia.* Chapel Hill: University of North Carolina Press for the Institute of Early American History and Culture, Williamsburg, VA, 1955.

Jefferson, T. (1784b). *Notes on the establishment of a money unit and of a coinage for the United States* (handwritten manuscript). Washington, DC: Library of Congress.

Jefferson, T. (1785). *Notes on the establishment of a money unit and of a coinage for the United States.* Paris, France: Author. (The notes are reproduced in P. F. Ford (Ed.), *The works of Thomas Jefferson* (Vol. 4, pp. 297–313). New York, NY: G. P Putnam's Sons. This was published in 1904).

Jefferson, Thomas to Edward Carrington, May 27, 1788 (in H. A. Washington (Ed.), *The writings of Thomas Jefferson*, New York, NY: H. W. Derby, 1861).

Jess, Z. (1799). *The American tutor's assistant ...* Wilmington, DE: Bonsal and Niles.

Kamens, D. H., & Benavot, A. (1991). Elite knowledge for the masses: The origins and spread of mathematics and science in national curricula. *American Journal of Education, 99*(2), 137–180.

Karpinski, L. C. (1980). *Bibliography of mathematical works printed in America through 1850.* New York, NY: Arno Press.

Keraliya, R. A., & Patel, M. M. (2014). Effect of viscosity of hyprophilic coating, polymer on lag time of atendolol pulsatile press coated tablets. *Journal of Pharmaceutical Chemistry, 1*(1), 1–9.

Kersey, J. (1683). *Mr Wingate's Arithmetick, containing a plain and familiar method for attaining the knowledge and practice of common arithmetick (8th ed.).* London, UK: John Martyn.

Kilpatrick, J., & Izsák, A. (2008). A history of algebra in the school curriculum. In C. E. Greenes & R. Rubenstein (Eds.), *Algebra and algebraic thinking in school mathematics: Seventieth yearbook* (pp. 3–18). Reston, VA: National Council of Teachers of Mathematics.

Kleeberg, J. M. (1992). The New Yorke in America token. In J. M. Kleeberg (Ed.), *Money of pre-Federal America* (pp. 15–57). Proceedings of the Coinage of the Americas Conference, held at the American Numismatic Society May 4, 1991, in New York.

Lee, C. (1797). *The American accountant; being a plain, practical and systematic compendium of Federal arithmetic* ... Lansingburgh, NY: William W. Wands.

Lewin, C. G. (1970). An early book on compound interest—Richard Witt's arithmeticall questions. *Journal of the Institute of Actuaries, 96*(1), 121–132.

Lewin, C. G. (1981). Compound interest in the seventeenth century. *Journal of the Institute of Actuaries, 108*(3), 423–442

Leybourn, W. (1690). *Cursus mathematicus.* London, UK: Author.

Linklater, A. (2003). *Measuring America: How the United States was shaped by the greatest land sale in history.* New York, NY: Plume.

Littlefield, G. E. (1904). *Early schools and school-books of New England.* Boston, MA: The Club of Odd Volumes.

Lossing, B. J. (1851). *The pictorial field-book of the Revolution.* New York, NY: Harper & Brothers.

Lutz, A. (1929). *Emma Willard: Daughter of democracy.* Boston, MA: Houghton Mifflin.

Macintyre, S., & Clark, A. (2004). *The history wars.* Melbourne, Australia: Melbourne University Press.

Madison, J. (1865). *Letters and other writings of James Monroe 1769–1793.* Philadelphia, PA: J. B. Lippincott & Co.

Maier, P. (2010). *Ratification: The people debate the constitution, 1787–1788.* New York, NY: Simon & Schuster.

Martin, G. H. (1897). *The evolution of the Massachusetts public school system: A historical sketch.* New York, NY: D. Appleton and Company.

Massey, J. E. (1976). Early money substitutes. In E. Newman & R. Doty (Eds.), *Studies on money in early America* (pp. 15–24). New York, NY: American Numismatic Society.

McCusker , J. J. (1992). *Money and exchange in Europe and America 1600–1775,* Chapel Hill, NC: University of North Carolina Press.

McCusker, J. J. (2001). *How much is that in real money? A historical commodity price index for use as a deflator of money values in the economy of the United States,* New Castle, DE: Oak Knoll Press.

Mehl, R. M. (1933). *The Star rare coin encyclopedia and premium catalog.* Fort Worth, TX: Numismatic Co. of Texas.

Michael, I. (1993). The textbook as a commodity: Walkingame's *The Tutor's Assistant. Paradigm, 12,* 2–10.

Miter (1896, September 8). Our London letter. *The American Stationer, 40*(10), 367–368 and 379.

Monaghan, E. J. (2007). *Learning to read and write in colonial America.* Amherst, MA: University of Massachusetts Press.

Money, J. (1993). Teaching in the market place, or *"Caesar adsum jam forte Pompey aderat":* The retaining of knowledge in provincial England during the 18th century. In J. Brewer & R. Porter (Eds.), *Consumption and the world of goods* (pp. 335–377). London, UK: Routledge.

Monroe, W. S. (1917). *Development of arithmetic as a school subject.* Washington, DC: Government Printing Office.

Morison, S. E. (1956). *The intellectual life of colonial New England*. Ithaca, NY: New York University Press.

Morris, R. (1782, January 15). Robert Morris to the President of Congress, January 15, 1782. In J. Boyd (Ed.), *The papers of Thomas Jefferson 16, March 1784 to February 1785* (pp. 160–169). Princeton, NJ: Princeton University Press.

Mouton, G. (1670). *Observationes diametrorum solis et lunae apparentium, meridianarumque aliquot altitudinum, cum tabula declinationum solis; Dissertatio de dierum naturalium inaequalitate,...* Lyons, France.

Napier, J. (1614). *Mirifici logarithmorum canonis descriptio*. Edinburgh, UK: Andrew Hart.

Napier, J. (1619). *The wonderful canon of logarithms*. Edinburgh, UK: William Home Lizars, 1857 [English translation by Herschell Filipowski]

Newman, E. P. (1990). *The early paper money of America*. Fairfield, OH: Krause Publications.

Nishikawa, S. (1987). The economy of Chōshū on the eve of industrialization. *The Economic Studies Quarterly, 38*(4), 209–222.

Notes on the Early History of the Mint. (1867). *Historical Magazine, 1*.

O'Brien, C. C. (1996). Thomas Jefferson: Radical and racist. *The Atlantic Monthly, 278*(4), 53–74.

Ogburn, W. F. (1922). *Social changes with respect to culture and original nature*. New York, NY: B. W. Huebsch.

Ogg, F. A. (1927). *Builders of the Republic*. New Haven, NJ: Yale University Press.

Patterson, R. (1819). *A treatise of practical arithmetic intended for the use of schools* (2 parts). Pittsburgh, PA: R. Patterson and Lambdin.

Peden, W. (Ed.). (1955). *Notes on the State of Virginia by Thomas Jefferson*. Chapel Hill, NC: University of North Carolina Press.

Perlmann, J., & Margo, R. (2001). *Women's work? American schoolteachers, 1650–1920*. Chicago, IL: University of Chicago Press.

Pike, N. (1788). *A new and complete system of arithmetic, composed for the use of the citizens of the United States*. Newburyport, MA: John Mycall.

Pike, N. (1793). *Abridgement for the new and complete system on arithmetic, composed for the use and adapted to the commerce of the citizens of the United States*. Newburyport, MA: Isaiah Thomas.

Pike, N. (1797). *A new and complete system of arithmetic, composed for the use of the citizens of the United States* (2nd ed.). Worcester, MA: John Mycall.

Pike, S. (1822). *The teacher's assistant or a system of practical arithmetic; wherein the several rules of that useful science, are illustrated by a variety of examples, a large proportion of which are in Federal money. The whole is designed to abridge the labour of teachers, and to facilitate the instruction of youth*. Philadelphia, PA: Benjamin Warner.

Plimpton, G. A. (1916). *The hornbook and its use in America*. Worcester, MA: American Antiquarian Society.

Putnam, E. (1885). A Salem dame-school. *The Atlantic Monthly, 55*(327), 53–58.

Radford, L. (2003). On the epistemological limits of language: Mathematical knowledge and social practice during the renaissance. *Educational Studies in Mathematics, 52*(2), 123–150.

Rappleye, C. (2010). *Robert Morris: Financier of the American Revolution*. New York, NY: Simon & Schuster.

Rashed, R. (1994). *The development of Arabic mathematics: Between arithmetic and algebra*. Dordrecht, The Netherlands: Kluwer.

Raven, J. (1997). The export of books to colonial North America. *Publishing History, 42*, 11–49.

Reid, J. P. (1991). *Constitutional history of the American Revolution, III: The authority to legislate*. Madison, WN: University of Wisconsin Press.

Ring, B. (1993). *Girlhood embroidery: American samplers and pictorial needlework, 1650–1850*. New York, NY: Knopf Publishers.

Roach, J. (1971). *Public examinations in England 1850–1900*. Cambridge, UK: Cambridge University Press.

Robinson, H. (1870). *The progressive higher arithmetic for schools, academies, and mercantile colleges*. New York: Ivison, Blakeman, Taylor & Co.

Root, E. (1795). *An introduction to arithmetic for the use of common schools*. Norwich, CT: Thomas Hubbard.

Ryan, K. R., & Cooper, J. M. C. (2010). Colonial origins. In L. Mafrici (Ed.), *Those who can teach* (12th ed.). Wadsworth: Cengage Learning.

Seaman, W. H. (1902, March). How Uncle Sam got a decimal coinage. *School Science, 232–236*.

Seybolt, R. F. (1917). *Apprenticeship and apprenticeship education in colonial New England and New York*. New York, NY: Teachers College, Columbia University.

Seybolt, R. F. (1921). The evening schools of colonial New York City. In New York State Department of Education (Ed.), *Annual report of the State Department of Education* (pp. 630–652). Albany, NY: State Department of Education.

Seybolt, R. F. (1935). *The private schools of colonial Boston*. Cambridge, MA: Harvard University Press.

Simons, L. G. (1924). *Introduction of algebra into American schools in the 18th century*. Washington, DC: Department of the Interior Bureau of Education.

Simons, L. G. (1936). Short stories in colonial geometry. *Osiris, 1*, 584–605.

Sinclair, N. (2008). *The history of the geometry curriculum in the United States*. Charlotte, NC: Information Age Publishing, Inc.

Skemp, R. (1976). Instrumental understanding and relational understanding. *Mathematics Teaching, 77*, 20–26.

Smith, D. E., & Karpinski, L. C. (1911). *The Hindu-Arabic numerals*. Boston, MA: Ginn and Company.

Sosin, J. M. (1964). Imperial regulation of colonial paper money, 1764–1773. *Pennsylvania Magazine of History and Biography, 88*(2), 174–198.

Sprague, W. B., & Lathrop, L. E. (1857). Chauncey Lee, D.D. In W. B. Sprague, *Annals of the American pulpit: Trinitarian Congregational: Or commemorative notices of distinguished American clergymen of various denominations, from the early settlement of the country to the close of the year 1855, with historical introductions* (Vol. 2, pp. 288–291). New York, NY: Robert Carter and Brothers.

Stafford, M. H. (1941). *A genealogy of the Kidder family: Comprising the descendants in the male line of Ensign James Kidder, 1626-1676, or Cambridge and Billerica in the colony of Massachusetts Bay*. Rutland, VT: Tuttle Pub. Co.

Stedall, J. (2012). *The history of mathematics: A very short introduction.* Oxford, UK: Oxford University Press.

Sterry, C., & Sterry, J. (1790). *The American youth: Being a new and complete course of introductory mathematics, designed for the use of private students.* Providence, RI: Authors.

Sterry, C., & Sterry, J. (1795). *A complete exercise book in arithmetic, designed for the use of schools in the United States.* Norwich, CT: John Sterry & Co.

Stevin, S. (1585). *De Thiende.* Leyden, The Netherlands: The University of Leyden.

Swan, S. B. (1977). *American women and their needlework 1700–1850.* New York, NY: Holt, Rinehart and Winston.

Swetz, F. (1987). *Capitalism and arithmetic: The new math of the 15th century.* La Salle, IL: Open Court. .

Tabak, J. (2004). *Numbers: Computers, philosophers, and the search for meaning.* New York, NY: Facts on File.

Tharp, P. (1798). *A new and complete system of Federal arithmetic.* Newburgh, NY: D. Denniston.

Thomas, I. & Andrews, E. J. (1809). *Preface to the third octavo edition of A new and complete edition of arithmetick, composed for the use of citizens of the United States: Seventh edition for the use of schools.* Boston, MA: Author.

Thomson, J. B. (1874). *Unification of weights and measures, the metric system: Its claims as an international standard of metrology.* New York, NY: Clark & Maynard.

Todd, J., Jess, Z., Waring, W., & Paul, J. (1800). *The American tutor's assistant; or, a compendious system of practical arithmetic; ...* Philadelphia, PA: Zachariah Polson, Junior.

Tuer, A. M. (1896). *History of the horn-book.* London, UK: Leadenhall Press.

Urban, W. J., & Wagoner, J. L. (1996), *American education: A history.* New York, NY: McGraw-Hill.

U.S. Department of Commerce, Bureau of the Census (1975). *Historical statistics of the United States: Colonial times to 1970.* Washington, DC: Government Printing Office.

Van Egmond, W. (1980). *Practical mathematics in the Italian Renaissance: A catalog of Italian abbacus manuscripts and printed books to 1600.* Firenze, Italy: Istituto E Museo di Storia Della Scienza.

Venema, P. (1730). *Arithmetica of Cyffer-Konst, volgens de Munten Maten en Gewigten te Nieu-York, gebruykelyk als mede een kort Ontwerp van de Algebra.* New York, NY: Jacob Goelet.

Viète, F. (1579). *Canon mathematicus seu ad triangula cum appendicibus.* Paris, France: Jean Mettayer.

Vignery, J. R. (1966). *The French Revolution and the schools.* Madison, WI: University of Wisconsin.

Walkingame, F. (1785). *The tutor's assistant: Being a compendium of arithmetic and a complete question book* (21st ed.). London, UK: J. Scratcherd & I. Whitaker.

Walsh, M. (1801). *A new system of mercantile arithmetic: Adapted to the commerce of the United States, in its domestic and foreign relations; with forms of accounts, and other writings usually occurring in trade.* Newburyport, MA: Edmund M. Blunt.

Wardhaugh, B. (2012). *Poor Robyn's prophecies: A curious almanac, and the everyday mathematics of Georgian Britain.* Oxford, UK: Oxford University Press.

Webster, N. (1789). *Dissertations on the English* language: *With notes, historical and critical, to which is added, by way of appendix, an essay on a reformed mode of spelling, with Dr. Franklin's arguments on that subject.* Boston, MA: Isaiah Thomas.

Westbury, I. (1980). Change and stability in the curriculum: An overview of the questions. In H. G. Steiner (Ed.), *Comparative studies of mathematics curricula: Change and stability 1960–1980* (pp. 12–36). Bielefeld, Germany: Institut für Didaktik der Mathematik-Universität Bielefeld.

Wickersham, J. P. (1886). *A history of education in Pennsylvania.* Lancaster, PA: Inquirer Publishing Company.

Wiencek, H. (2012). *Master of the mountain: Thomas Jefferson and his slaves.* New York, NY: Farrar, Strauss and Giroux.

Wilkins, J. (1668). *Essay towards a real character and a philosophical language.* Held in archives of Cambridge University.

Wilson, D. L. (1992). Thomas Jefferson and the character issue. *The Atlantic Monthly, 270*(3), 37–74.

Wilson, D. L. (1996, October). Counter points. *The Atlantic Monthly* online.

Windschuttle, K. (1996). *The killing of history: How literary critics and social theorists are murdering our past.* San Francisco, CA: Encounter Books.

Wingate, E. (1624). *L'usage de la règle de proportion en arithmétique.* Paris, France: Author.

Wingate, E. (1630). *Of natural and artificiall arithmetique.* London, United Kingdom: Author.

Workman, B. (1789). *The American accountant or schoolmaster's new assistant* ... Philadelphia, PA: John M'Culloch.

Workman, B. (1793). *The American accountant or schoolmaster's new assistant ... Revised and corrected by Robert Patterson.* Philadelphia, PA: W. Young.

"Wright on measuring the meridian—Wright, Wren and Wilkins on an universal measure." (1805). *The Philosophical Magazine, 21,* 163–173

Wu, D. T. (1992). The time lag in nucleation theory. *The Journal of Chemical Physics, 97,* 2644.

Author Index

M. A. (Ken) Clements, & N. F. Ellerton, *Thomas Jefferson and his decimals 1775–1810: Neglected years
in the history of U.S. school mathematics*, DOI 10.1007/978-3-319-02505-6,
© Springer International Publishing Switzerland 2015

Subject Index

M. A. (Ken) Clements, & N. F. Ellerton, *Thomas Jefferson and his decimals 1775–1810: Neglected years
in the history of U.S. school mathematics*, DOI 10.1007/978-3-319-02505-6,
© Springer International Publishing Switzerland 2015